US Environmental History

US Environmental History

Inviting Doomsday

John Wills

EDINBURGH
University Press

© John Wills, 2013

Edinburgh University Press Ltd
22 George Square, Edinburgh EH8 9LF
www.euppublishing.com

Typeset in 11/12.5 Sabon by
Servis Filmsetting Ltd, Stockport, Cheshire, and
printed and bound in Great Britain by
CPI Group (UK) Ltd, Croydon CR0 4YY

A CIP record for this book is available from the British Library

ISBN 978 0 7486 2263 4 (hardback)
ISBN 978 0 7486 2264 1(paperback)
ISBN 978 0 7486 2979 4 (webready PDF)
ISBN 978 0 7486 7236 3 (epub)
ISBN 978 0 7486 7235 6 (Amazon ebook)

The right of John Wills to be identified as author of this work has
been asserted in accordance with the Copyright, Designs and
Patents Act 1988.

Published with the support of the Edinburgh University Scholarly
Publishing Initiatives Fund.

Contents

Figures

Introduction

JACKSON CURTIS. I was listening to the broadcast and I was wondering what is exactly that's gonna start in Hollywood.
CHARLIE FROST. It's the apocalypse. End of days. The Judgment Day, the end of the world, my friend.

2012 (2009)

In 2009, Hollywood provided a sneak preview of 'the apocalypse. End of Days'. With Western interpretations of the Mayan prophecy serving as inspiration, the blockbuster movie *2012*, directed by Roland Emmerich, mapped out the contours of the forthcoming Judgement Day. As predicted in the ancient Mayan calendar, the world would end on 21 December 2012. The film's tag line read simply, 'We were warned'.

2012 homed in on the trials and tribulations of failed novelist and unlikely action hero Jackson Curtis (John Cusack) and his desperate attempts to circumvent Armageddon. With the warming of the Earth's core triggering the falling apart of the United States, with whole landscapes disappearing, Curtis and his dysfunctional family attempt escape by commandeering limousine, Winnebago, and private jet. Watching over his crumbling capitalist empire, reflecting on his role as the very last commander-in-chief, President Thomas Wilson (Danny Glover) speaks to the nation: 'My fellow Americans. This will be the last time I address you. As you know, catastrophe has struck our nation . . . has struck the world. I wish I

Figure I.1 *2012* (Columbia Pictures, 2009).

could tell you we can prevent the coming destruction. We cannot.' Iconic monuments of America fall before the waves of mass destruction. Washington Monument, the Golden Gate Bridge, and the White House all vanish. As the Yellowstone caldera erupts, conspiracy theorist Charlie Frost (Woody Harrelson) shouts, 'This marks the last day of the United States of America. And, by tomorrow, all of mankind.'

2012 was designed to be the ultimate doomsday experience. To do justice to the end of the world, director Roland Emmerich crafted a multi-disaster adventure, melding together scenes from *The Poseidon Adventure* (1972), *Titanic* (1997), *Volcano* (1997), *The Towering Inferno* (1974) and more. The celluloid commixture fashioned catastrophe into great entertainment. *USA Today* dubbed *2012* 'the mother of all disaster movies'. Emmerich visualised a range of classic doomsday scenarios to help give the Mayan prophecy realism and depth: Los Angeles collapsing in on itself by earthquakes; Yellowstone erupting by its giant volcano; New York swept away by tsunami; Las Vegas Strip folding like a pack of cards in the desert. It tapped Cold War and superpower themes to foster danger. Characters included a despicable Russian oligarch and a menacing Chinese military force. To increase the impact of the film, Emmerich tapped as many American concerns as possible.[1]

At its core, *2012* was a product of American environmental anxiety, a Tinseltown reflection on unfolding environmental crisis in twenty-first-century USA. Outside Hollywood, two real environmental disasters, Hurricane Katrina and the *Deepwater Horizon* oil spill, played out, rallying concern. Evidence of global warming, forwarded in the popular documentary *An Inconvenient Truth* (2006) by Al Gore, highlighted imminent danger. And yet, the US government had a poor record in tackling environmental issues, its failure to sign the Kyoto climate accords a signal of stewardship irresponsibility. The United States seemed to be inviting doomsday while also entertaining it. The exchange between Curtis and Frost, over Hollywood and Armageddon, spoke to this contemporary condition. The film also offered (unlikely) American triumph in the theatre of impending environmental disaster.

The United States has been inviting doomsday for quite some time. Disaster is not an invention of Hollywood, Al Gore or environmental groups such as the Sierra Club. In fact, the invitation dates back several hundred years. Recent environmental disasters, such as Hurricane Katrina, are rooted in the past. A broader trajectory is involved. American environmental history is a history of disaster. A complexity of factors has determined the downward spiral: the early colonial mindset, the capitalist system, white-coat scientists and technology, the everyday consumer, myths of abundance, religious dominion, and manifest destiny. Problematic notions of the technological fix (to all environmental problems), a common anthropocentric way of conceiving the world, and the classifying of nature as a simple resource have furthered the doomsday condition. The inviting of doomsday is the historic collision of science and technology, politics and religion, capitalism and catastrophe culture, and problematic attitudes toward the natural world.

American environmental disasters help us to understand this doomsday condition. They include the San Francisco Fire (1906), Three Mile Island nuclear accident (1979), and more. In this book, I focus on a small selection of American disasters spread across a broad time frame. The book is an exploration of a range of catastrophe-based case studies that chart an American trajectory towards environmental ruin. Some of the doomsday scenarios are real and significant, while others are fictive and imagined. As in the movie *2012*, film has shaped our imaginations, our fears, but equally popular imaginations of doomsday go back much further. And their expressions and outputs are far more diverse than the Hollywood film machine. Gathered together, the examples herein highlight an enduring and dangerous environmental attitude (of destructiveness), and a culture of environmental fear (which I label 'catastrophe culture') in the United States. The book also shows how little has been learnt through these disasters, and that owing to this, America is still very much inviting doomsday in the twenty-first century, in 2012.

The book is split into four sections. The prologue introduces the 'doomsday invitation' at Jamestown at the beginning of the colonial period. Here, Americans first embark on the doomsday project. Caught up in fears over wilderness, plague and bestiality, of nature heralding doom, New Englanders introduce all kinds of changes and contagions of their own into the American landscape. The second section tackles the 'age of massacre', the nineteenth century. Here, killing is commonplace, and disaster very much the work of humankind: the massacring of bison and passenger pigeons, alongside Native Americans and Civil War soldiers indicative of a new scale of environmental ruin. Deathly images of 'animal-cide' and wilderness lost are dominant. Doomsday is clearly manufactured. The third section covers post-1945 technological threats and dangers in the form of nuclear (Nevada Test Site and Doom Town), chemical (DDT and *Silent Spring*), and oil (Santa Barbara and *Deepwater Horizon* spills). Images here include the radiated world, a chemically poisoned town, and a return to barbarism in the absence of oil. The final section explores Hurricane Katrina, global warming, and the end of both city and nature. Doomsday images seem more complicated here, with the unravelling of historic codes of nature and civilisation occurring.

The four sections, this series of disasters, chart a trajectory towards environmental collapse. The doomsday scenarios are first seen to be nature-made (the wilderness), then man-made (through extinctions), then technologically made (chemicals, the bomb), then a fusion of all (Katrina blamed on global warming, weather systems and urban design). They amount to key stages of doomsday. They implicate the workings of first nature, then humankind and then technology in 'inviting doomsday'. Notably, the potential for a true 'end of world' scenario increases across time. The shift from disasters first caused by nature to later, a synthesis of nature, humanity and technology creates a spiral in terms of damage, until images of nuclear doomsday and global warming threaten total collapse.

How to interpret this trajectory is one initial challenge. These four stages might represent stages of decline: the end of the wilderness environment, the death of animals and humans, the fall of towns and cities, and ultimately the demise of the global environment. Disaster also displays a variant scope and trajectory across time: initially, it is seen as being intimate and one-directional (disaster emerging from the wilderness experience), then by direct exchange (human on human, gun on animal), then indirect (pollution and pesticides, such as DDT), then distant (deriving from global warming). Also, the four stages might suggest the gradual failure of life support: the natural environment, human resources, chemical technology and even artificial means all, in turn, proving incapable of supporting life on Earth. Or the stages might be seen as part of an ongoing battle for resources, a battle over the American energy landscape, with disasters as distinct episodes of this conflict. Finally, and most radically, the four stages might compare with the Four Horsemen of the Apocalypse: conquest (of the New World), war (against humanity/nature on the nineteenth-century frontier), famine (the challenge of limited resources in the twentieth century) and death from all. Other elements within these doomsday scenarios will be investigated, such as the role of media in environmental perception, the social construction of disaster, and the distinctive catastrophe culture that surrounds environmental ruin. All will be explored within *Inviting Doomsday*.

This project was funded by the British Academy. The Historical Society of Southern California aided in the project on the Santa Barbara oil spill. Archives used include the Department of Special Collections at the University Library, University of California, Santa Barbara, the Nuclear Testing Archive at Las Vegas, the Beinecke Library at Yale University, the Huntingdon Library at Pasadena, the Gene Autry Museum of the American West and the Braun Research Library in Los Angeles, and the British Library in London. Archivists Martha deMarre at Vegas, Marva Felchin at the Autry and Liza Posas at the Braun all proved wonderful with their assistance. Janet Wasko, Ralph Lutts, Joe Street and Malcolm McLaughlin helped refine individual chapters. Nicola Ramsey supported the project throughout. I also thank staff and students at the School of History and Centre for American Studies at Kent University, colleagues Kenneth Fincham, Will Pettigrew and Claire Taylor, and the fantastic Claire Wenham.

Note

1. Claudia Puig, '*2012*: Now that's Armageddon', *USA Today*, 13 November 2009.

Prologue: The Invitation

Back in the 1600s, the 'New World' of America furthered the expansion of the Old World. Caught within the colonial gaze of the period, dignitaries and explorers alike envisaged America as a realm to be moulded in the European image. They categorised the mystery territory as a prize to be had in the competitive theatre of global expansion. Colonial boosters hoped the country might facilitate fresh trade routes and furnish vast mineral wealth. Settlers imagined an old and familiar rural landscape made anew. Trade, Christianity and empire coalesced, and together forwarded an experimental takeover of new land. For individual colonists, boarding galleons about to travel the Atlantic in the early seventeenth century, 'New America' came to embody many positive Old World attributes: property, fortune, and a way to impress the Crown. America invited speculation, excitement and conquest. But with the advent of New World conquest came a hidden invitation. Among the hubris of colonial chatter, the charters of European nations, and the kegs of salted meat and gunpowder, each vessel carried something darker and more sinister. The colonists departing for American shores had invited doomsday on board. Caught within the expansive schemes for new America were the seeds of environmental doomsday. Disaster could be found in the hopes and fears of European colonists, and also in their plans.

The colonial mindset cast America as a doomsday landscape long before arrival or development. On board the *Susan Constant*, the *Discovery* and the *Godspeed* heading for New England, British entrepreneurs looked on American life with fear and angst before reaching dry land. Alongside the official colonial gaze of acquisition, early colonists brought with them a doomsday gaze, an early form of catastrophe culture. Folklore and ignorance forged American soil into a realm of evil wilderness, misery and death. Avid Christians drew on biblical notions of the Apocalypse set out in Revelation, fearing the potential for extreme darkness ahead on the new shores of America. Literature and theatre of the day cast doubts over the colonial enterprise. In *The Tempest* (1611), William Shakespeare constructed the island on which Prospero lands as a symbolic landscape of the New World, and employed the character of Caliban as an avatar of wild America, oft depicted as a deformed and frightening half fish, half man, born of Sycorax and the devil. A 'metaphor for English expansion into America', *The Tempest* told of the potential horror of the New World. Tommaso Campanella's *The City of the Sun* (1602) and later Jonathan Swift's *Gulliver's Travels* (1726) equally imagined doomed experiences on New World lands. In *New Atlantis* (1627), Francis Bacon relayed the fear of journeying to the New World, of 'finding ourselves, in the midst of the greatest wilderness of waters in the world, without victuals, we gave ourselves for lost men and prepared for death'. For Bacon, only the 'lucky' few made it safely to the New Atlantis. Popular disaster imaginings of mass culture included sailor's stories of being lost at sea, as well as sea creatures patrolling New World maps, highlighting the hidden peril of colonial expansion. A sense of sacrifice and folly accompanied the New World experiment, raising the spectre that things could go wrong as often as right.[1]

Colonists brought with them detailed plans for the New World. Europeans typically travelled with royal authorisation for their territorial claims. Landing at Chesapeake Bay in April 1607, the London Company (Virginia Company) carried with it the right to a significant stretch of America as decreed by King James I. The first Virginia Charter of 1606 granted

> Licence, to make Habitation, Plantation, and to deduce a Colony of sundry of our People into that Part of *America*, commonly called VIRGINIA, and other Parts and Territories in *America*, either appertaining unto us, or which are not now actually possessed by any *Christian* Prince or People.

Forwarding plans to set up two plantations on the eastern coast, the Virginia Charter organised and guided the colonial endeavour. Along with granting legitimacy in global affairs, the charter scripted the New World story and summed up broader intentions for America. It also marked out an environmental trajectory for the New World, an ecological modus operandi. In the case of the Virginia Company, a charter for colonisation served as an invitation for doomsday.

The Virginia Charter of 1606 set a number of important precedents for navigating the American environment. Firstly, the colonial doomsday gaze classified all American land as idle, free and ripe for wholesale exploitation. Colonists regarded Native American agriculture and hunting as inconsequential and irrelevant. Only Europeans knew how to make best use of the land. Thomas More argued in *Utopia* (1516) how idle land existed for the remaking. The colonial gaze cast American nature as a simple resource, there to be taken. The charter decreed that, on arrival, company men 'shall have all the Lands, Woods, Soil, Grounds, Havens, Ports, Rivers, Mines, Minerals, Marshes, Waters, Fishings, Commodities, and Hereditaments, whatsoever'. Nature was disassembled into discrete parts (or products) to take and use. Fashionable scientific concepts of the natural world as eminently reducible and understandable influenced this approach. As an Enlightenment scholar, Francis Bacon likened nature to a machine, something dissectible, easy to comprehend and full of potential to work for humankind. In stark contrast to traditional European folklore that projected a mysterious evil wilderness, Baconian nature was neutral and empty. But Bacon still 'othered' nature and justified its conquest. Doomsday was rooted as much in these hopes (and plans) of English colonists, as it was in disaster imaginings of being overwhelmed by barbarism and wilderness on new shores. Built into the making of the New Atlantis were significant environmental costs. To build Utopia, to make New England, required the wiping out of old nature first, a veritable environmental doomsday project.

Along with a utilitarian and enlightened quest to remake the land, the charter also presented the colonial experiment as divinely authorised and directed. The Bible justified environmental stewardship and land development. The charter extended the message of God to the New World, and situated the overcoming of wilderness and the taming of 'the savage' as a high priority. James I ordered the 'propagating of Christian Religion to such People, as yet live in Darkness and miserable Ignorance of the true Knowledge and Worship of God, and may in time bring the Infidels and Savages, living in those parts, to human Civility'. The old environmental world faced extinguishment as it clashed with European religious doctrine.

Added to that, the charter pushed a hard trade agenda. God combined with money. The Virginia Company first and foremost represented a mercantile venture, an investment gamble of 1606. The charter sanctioned its members 'to dig, mine, and search for all Manner of Mines of Gold, Silver, and Copper, as well within any part of their said several Colonies, as for the said main Lands'. New America was about profit accumulation. Travellers to the New World were merchants and gentleman, not settlers or farmers. Profits were meant to head home, to England, 'YIELDING therefore, to Us, our Heirs and Successors, the fifth Part only of all the same Gold and Silver, and the fifteenth Part of all the same Copper'. Unfortunately, returns proved to be not so forthcoming. The third Virginia Charter (1612) even established a lottery to help fund the colonial endeavour.

The Virginia Charter thus served notice on the environment. It represented a legal charter for environmental domination. The document articulated the operation of nature for the benefit of Europeans, a religious sanction for colonisation, a trade impulse to seize resources, and a stark ignorance of the environmental reality of a new American landscape. By its dedication to tap resources and reshape the land, the charter forwarded a persuasive line in doomsday logic. Journalist Bill McKibben claimed that wild and untouched nature died in America in the late twentieth century thanks to modern pollution and global environmental problems. But nature actually ended on the eastern seaboard in the early 1600s.[2]

On arrival at Jamestown and Plymouth colonies, the Europeans introduced all kinds of changes to the New World. Merchants unloaded domestic farm breeds and English plants and crops, all set to uproot native soil and resident fauna. They carried the awesome technology of muskets and gunpowder, ready to wage war against the wilderness environment. Traders unknowingly carried foreign pathogens to kill off native peoples. The ships amounted to contagion vessels, doomsday sails. For historian Alfred Crosbie, such vessels were 'like giant viruses fastening to the sides of [a] gigantic bacterium and injecting into it their DNA, usurping its internal processes for their own purposes'. The colonial experiment amounted to 'ecological imperialism'. Allied to this, a doomsday mindset, an attitude of destructiveness that cast native peoples and animals as eminently expendable. As environmental historian John Opie noted, colonists exhibited 'a passion for conquest'. Problems quickly emerged.[3]

In 1584, Sir Walter Raleigh had established a settlement at Roanoke on the eastern coast. Despite native assistance, Roanoake mysteriously disappeared, probably due to mass starvation or Indian war. A doomed and lost colony, Roanoke symbolised the real dangers of New World endeavour. At Jamestown in the early 1600s, London Company merchants struggled to survive. Faced with food shortages, crop failure, poor weather, Indians, malaria, mosquitoes, and a seemingly endless wilderness, the group of English gentlemen found themselves hopelessly overwhelmed. Most thought that their lives were doomed during the 'starving time' of 1609–10. Only sixty out of five hundred colonists survived the initial challenges of the eastern seaboard. At the Plymouth plantation, meanwhile, William Bradford recorded the 'hideous and desolate wilderness' engulfing him 'full of wild beasts and wild men'. As scholar Kevin Rozario highlighted, Bradford 'organized his history of settlement around a series of encounters with disasters'. Environmental historian Roderick Nash detailed how 'imagination multiplied fears' in this new wilderness climate. Catastrophe culture took hold, a doomsday vision imposed on the New World. Nature seemed very capable of destroying colonial life, of wiping it out. Settler John Hammond bemoaned how 'The Country is reported to be an unhealthy place, a nest of Rogues, whores, desolate and rooking persons; a place of intolerable labour, bad usage and hard Diet.' The darker vision of America seemed likely to win out, given the death narrative at Roanoke, then early Jamestown, America's first ground zeros.[4]

Yet, in the words of John Hammond, determined 'to bud forth, to spread further, to gather wealth', the colonial experiment gradually took root. New laws encouraged the planting of food, the nurturing of pigs and cattle. Over the course of several decades, merchants transformed the Old World on the eastern seaboard into a familiar English monocultural range of clearings, farms and townships. In the process, the decay and death transferred to infect native lands and peoples, and a grander doomsday scenario began to unfold. As John Opie related, the displacement of native systems with European ones promoted 'ecosystem collapse'. The dooms-day vision was successfully transplanted. Smoke and death over Roanoke and Jamestown seemed minor compared to the wildfire about to spread to other parts of the nation.[5]

Notes

1. Gerald Graff and James Phelan, eds, *The Tempest* (Basingstoke: Macmillan, 2000), p. 146.
2. See Bill McKibben, *The End of Nature* (London: Viking, 1990).
3. Alfred Crosbie, *Ecological Imperialism: The Biological Expansion of Europe, 900–1900*, New Edition (Cambridge: Cambridge University Press, 2004), p. 227; John Opie, *Nature's Nation: An Environmental History of the United States* (Fort Worth: Harcourt Brace, 1998), p. 41.
4. Roderick Nash, *Wilderness and the American Mind* 3rd edition (New Haven: Yale University Press, 1982 [1967]), pp. 23–4, 26; Kevin Rozario, *The Culture of Calamity: Disaster and the Making of Modern America* (Chicago: University of Chicago Press, 2007), p. 35; John Hammond, *Leah and Rachel* (1656) pamphlet.
5. Hammond (1656); Opie (1998), p. 73.

CHAPTER ONE

Killing in the Wilderness

The American Civil War (1861–65) was about slavery, democracy, and nationhood. The Civil War was also a battle over the environment: a fight for the future of the national landscape, pitting 'industrial' against 'agrarian', 'city' versus 'rural', the 'machine' versus the 'seed crop', and 'modernity' versus 'traditionalism'. It was a battle over two conflicting landscape designs. The American South embodied a traditional agrarian system. Southern plantation owners exploited soil and people alike, nature and black Americans co-opted into a system of servitude. The products of service, tobacco and rice, generated the riches of the South. Monoculture in agriculture furnished a veritable Utopia for the slave barons; a barren land of depleted labour and infertile soil was the consequence. The North embraced a different kind of landscape: the industrial cityscape. Again, business magnates prospered, this time with legions of the working class caught up in the polluted fallout of the chimney-stack economy. The industrial machine clearly fired towards a different future, that of urban prosperity.

The battle over landscape also revealed a deeper, more troubled environmental narrative. The Civil War marked not just a fork in the road between an agricultural or urban future but contributed to America's march towards environmental collapse. In the colonial period, settlers and merchants often felt overwhelmed by their environment and imagined doomsday as the product of unforeseen plague or something evil in the wilderness. By the mid-nineteenth century, Americans exercised significant control over the continental expanse and exhibited huge destructive capabilities. Fear waned of losing civilisation in the wilderness. In its place, a disaster scenario, based around massacre and killing, arose. The spectre of doom shifted from nature's malice to humankind's own powers. The Civil War showed this important transition stage; it highlighted what people were now capable of. The conflict entailed the decimation of lives and land on an unbridled scale. It showed a new attitude of annihilation and set the tone for future environmental engagement. It marked the ascendency of doomsday culture.

The American environment provided the fuel for battle, the resources that made conflict possible. In the 1860s an energy landscape specific to wartime emerged, servicing a modern doomsday war machine. Men relied on rations such as navy beans, salted beef, and coffee beans that came from distant food contractors. With the soldiers travelled herds of cattle ready for slaughter. Troops foraged for fruit and vegetables, learning the landscape as they passed through it, and pillaging it for their survival. Cavalry relied on horses that, in turn, required daily fresh pasture. The two armies fought over territory and supplies. They desperately protected the paths of supply trains, defending their individual energy landscapes from enemy attack. Troops became fuel for the war machine, fodder for the fight. Unlike in twenty-first-century technological warfare, Civil War soldiers survived in poor conditions and slept close to the ground. Carting rucksacks of provisions and arms, they marched miles each day. They engaged with the landscape in a total and destructive way.

For four years, the land served as the ground zero for conflict. Local economies collapsed, cities such as Vicksburg and Charleston lay in ruins, and farmland was remodelled into battlefield trenches and breastworks. Artillery blasted buildings, fires devastated forests, and infantry wrecked agricultural land. Thousands lost their lives to conflict and disease, and family lines disappeared. The war machine left whole swathes of country derelict. The American Civil War spawned a series of environmental disasters spread out across the South. As historian Lisa Brady related, 'campaigns in Georgia and the Carolinas left behind an awesome vista of destruction'. Most poignantly at Arlington, antebellum plantations became postbellum mass graves. The conflict showed the new capabilities of Americans for manufacturing doomsday.[1]

The Civil War fashioned horrific individual landscapes of death and decay. Battlefields signified places of fear, brutality, hardship, and indiscriminate slaughter. Troops moved from one doomed landscape to the next, living in permanent states of agitation, anxiety and fear. The Civil War implanted a lasting catastrophe culture in these men. Individual battles showed the dark potential of humanity. The Battle of the Wilderness, 5–6 May 1864, was one such example. Two huge armies fought each other like brawling bears in the woods of Virginia, providing a bloody example of what Americans could do to one another. The battle drew on older fears of doom, of evil lurking in the wild and nature's malevolence. Folkloric fears of the Wilderness combined with modern forms of brutality. Nature-made and man-made disaster imaginations combined. From this, the Union and Confederacy manufactured a heady doomsday landscape in the wilds of Virginia. Soldiers felt like they were seeing the end of the world, witnessing total disaster. The Battle of the Wilderness was a sign of a downward spiral in American culture, a transitional shift towards a new stage of doomsday, a new environmental attitude of mass destruction. It fed into other massacres of the period, of Native Americans, bison and passenger pigeons.

The Battle of the Wilderness

By 1864, the scale of destruction and death amassed in the Civil War had shocked both sides. The Union and the Confederacy entered the third year of conflict sporting a mixture of fatigue and dogged spirit. While well equipped and supplied, the huge Union Army lacked veterans. One southerner remarked how Union troops resembled 'pigeons for Lee's veterans to shoot out', easy targets, akin to defenceless passenger pigeons. The Union's other weakness was its inability to master the southern environment. As scholar Ted Steinberg related, the red clay soil of Virginia regularly upset the Army of the Potomac, especially in winter, when roads and tracks became almost impassable because of water and mud. The Army of Northern Virginia, meanwhile, boasted of tradition and experience, and fought mostly on home territory. Troops travelled familiar landscapes, knew the advantages (and disadvantages) of terrain, and fought directly for 'their country'. While nature seemed on the side of the South, however, the southern army struggled with resources. The numbers of recruits continued to fall while the numbers of the Potomac spiralled. The agrarian landscape suffered under conflict. With rations limited, and cornmeal a tedious foodstuff, soldiers and horses alike were left foraging in fields.[2]

Lieutenant General Ulysses S. Grant hoped to destroy the Confederate army in 1864 by throwing everything at them in Virginia. To succeed, the Union needed to face the army of General Robert E. Lee in an open and neutral setting but first had to cross the Rapidan river and pass through a place called the Wilderness.

Despite its title, the Wilderness was a man-made landscape. Mined for iron ore and quartz, and extensively logged for industry, from the 1710s the Wilderness functioned as a working slave landscape under the command of Alexander Spotswood, governor of Virginia. Smelting and logging dislodged remnant indigenous hunting, and the landscape became tied to new economic forces. A stagecoach route ran through the Wilderness, while a small spread of buildings marked the area.

With its resources spent by the early nineteenth century, southerners deserted the region. Disorder and decay set in. Wild nature recolonised. Laurel, dwarf oaks, and cedar saplings took root. Leading up to the Civil War, the Wilderness seemed caught in its own battle between civilisation and nature, of machine decay versus organic reboot. As plants took hold, a few old houses and clearings were all that remained of a slave realm. The resultant landscape, an uncomfortable mix of old and new, captured a sense of the greater South in flux. As General Horatio King related, 'Milton's "brush with frizzled hair implicit" describes the tangle of underbrush with which the floor of the forest was encumbered'. Soldiers of the Civil War entered a work in progress, a landscape of half-height trees and scrub. The 'tangle' was a direct result of human use, as Civil War soldier and writer Morris Schaff remarked, 'the present timber aspect is due entirely to the iron furnaces and their complete destruction of the first noble growth'. The Iron Age produced a wilderness-type landscape far more inaccessible than centuries past.[3]

The Virginia Wilderness was essentially the wrong kind of wilderness. The muddy swamps hardly amounted to the majesty of Yosemite Falls, rising hills hardly evincing images of El Capitan. As Congress set aside Yosemite as a state park (later national park) in 1864, the Virginia Wilderness was about to be sacrificed in war.

While Yosemite suggested a revision in views towards American wilderness and a new appreciation for the natural environment, the Battle of the Wilderness showed a different logic at work. The Virginia battle drew extensively on older folklore attached to wilderness. It showed a destructive attitude to nature, not a romantic one. The majority of soldiers drew on devilish imagery associated with the non-human. Historian Edward Steere described the Virginia lands as a 'dreary wasteland', with 'ugly scars' due to the 'robbery of subsoil'. 'Casting eternal shadows over stagnant pools and marsh creek bottoms, this brooding jungle not only inspired the name but imposed the conditions of combat in its gloomy depths,' Steere claimed. Older, troubling meanings of wilderness loomed large on the battlefield.[4]

Union leaders proved well aware of the practical dangers the Virginia Wilderness posed. The Army of the Potomac could ill afford to be stuck in its swamps. Chief-of-staff Major General Andrew Humphreys hoped that the army 'might move so far beyond the Rapidan the first day that it would be able to pass out of the Wilderness and turn, or partly turn, the right flank of Lee before a general engagement took place', and continued, 'I do not perceive that there is anything to induce the belief that General Grant intended or wished to fight a battle in the Wilderness'. Union leaders feared numerical, artillery and cavalry advantages would all disappear in the wild. One soldier wrote how Lee's 'numbers would be so magnified in effectiveness, and Grant's so neutralized, by the natural difficulties and terror of the woods'. While resting just north of the Rapidan, cavalry officer Robert Stoddart Robertson felt that 'rollicking pleasure must give place for visions of desolation and blood, and our winter carnival be supplemented by the wild carnival of death'. Glancing hesitantly towards the river, 'beyond which lay the confederate hosts ready to welcome us "with hospitable hands to bloody graves"', Robertson feared great bloodshed.[5]

On the morning of 4 May, the first and second divisions of the Union army crossed the Rapidan at Germanna and Elys Fords. With supply trains caught some distance behind, the Union army halted deep in the Wilderness, losing its momentum and tactical opportunity. On 5 May the entire force of the Potomac engaged with the Army of Northern Virginia. Around 120,000 troops from the North faced 65,000 from the South. Moses Kingsbury recorded in his diary: 'left camp early in the morning and marched to the wilderness and there commenest a fite and fought all day in the woodes with heavy lose on both sides'. Sudden gains in one part of the Wilderness were offset by sudden losses elsewhere. Men bludgeoned men, generals desperately sought advantage. By the end of the day, neither side had a decisive hold on the landscape.[6]

The Wilderness experience

Grant feared a battle in the Wilderness where the Confederacy would hold clear advantage. The messy and chaotic landscape quickly led to widespread disorientation on the Union side. Bushwhacking through the woods without an effective map or route, Ambrose Burnside's IX corps failed to join a planned offensive. One colonel misplaced his regiment entirely. Union commander John Sedgwick lost most of his fine beaver-cloth uniform in the tangled thickets. On exiting the field of battle, his regimental band played 'O ain't you glad to get out of the Wilderness'. Making their way through the woods, Union troops encountered a series of earthworks and defences held by Virginia infantry. Greys hid beneath tree trunks, avoided open conflict, and prevented any sustained Union assault. According to historian Gordon Rhea, 'the wilderness offered an ideal battlefield' for Lee's men, with the rebels holding a 'natural fortress' in the thickets. On 6 May, Moxley Sorrel's Confederate regiment utilised a new railway line constructed in the woods, attacking Winfield Scott Hancock's troops with devastating effect. A 'vast, weird, horrible slaughter pen' was the result.[7]

The Union rank and file related the full horror of the battle unfolding. A strong and fearless army just days prior, after a few hours in the Wilderness, 'men were pouring from the woods like frightened birds from a roost'. Fears of being caught, lost or killed in the Wilderness abounded. Positioned in the 140th New York Volunteers, Porter Farley detailed the 'tragic experience' of first contact with the rebels:

> The instant the regiment showed itself and before it was fairly out into the clearing a line of smoke puffed from the edge of the woods opposite and a volley of musketry sent its bullets into our ranks. Its effect was serious.

Nonetheless, with a 'wild cheer', 'down the slope we rushed; killed and wounded men plunging face forward to the ground'. Farley witnessed the terror of the terrain, where 'we found ourselves fighting with an almost unseen enemy but suffering all the time a continuous loss'. In a letter dated 9 May, Private Wilbur Fisk reflected on the 'wilderness of woe', where 'every bush and twig was cut and splintered by the leaden balls'. 'I doubt if a single tree could have been found that had not been pierced several times with bullets,' Fisk prophesied, his own clothing hit by enemy fire. Fisk spent one night digging breastworks and, on consecutive days, advanced across the same stretch of land where 'dead comrades lay on ground'. On reception of yet another rebel counter-attack, Fisk reported the 'considerable disorder and confusion in our hasty retreat, and the regiment was more or less broken up'. Demoralised at losing his men, Fisk lamented, 'I had been fighting to the best of my ability for Uncle Sam's Constitution, and now I thought it of about as much importance to me individually, to pay a little attention to my own.' Some 264 of his regiment died in the Wilderness. As one soldier related, 'the Federal troops have shown themselves less able than the Confederates to cope with the difficulties of forest fighting and more subject to its terrors'. The Union quickly fell into disunion.[8]

For the South, too, the Wilderness caused problems. 'Never have I fought on a more unsatisfactory battlefield,' declared one commander. Lieutenant Colonel William W. Swan agreed: 'A more difficult or disagreeable field of battle could not well be imagined.' On 6 May, a Union offensive along Orange Plank Road triggered havoc in southern ranks. Many troops fled the scene. In angry mood, Lee admonished the leader, 'General McGowan, is this splendid brigade of yours running like a flock of geese?'[9]

The Wilderness experience proved a significant challenge for both armies. The heat of the Virginia forest undermined morale. James George, a Union musician, moaned how the Wilderness was 'uncomfortably warm', to the extent that 'while the fighting was going I had my hair cut short'. Another soldier related the extreme heat of the situation: 'In the close woods the atmosphere is like that of an oven.' The density of the undergrowth proved a key obstacle to movement. Troops referred to the Wilderness as a 'jungle', a disorderly mass of vines and swampy land, an exotic, unfamiliar realm where nature ruled. Orderly marches ground to a standstill. Trees caused bullets to ricochet and deflect. With a sense of biblical damnation, one Confederate outlined being trapped by the slow pace, with his battalion 'as slow at getting into the Wilderness as the children of Israel were in getting out of one'.[10]

With few clearings, attempts to use artillery, the machines of war, fell fallow in the organic realm. Conceiving every machine as a musical instrument, George described how 'The trees were so thick they could not get artillery to play good at all.' The machine in the jungle struggled to operate. Elsewhere, swamps bogged down troop movement, one regiment 'impeded and bewildered' by their environment, and 'completely unnerved' when musket sounds drifted through. Corps rapidly lost any sense of organisation or alignment. Gaps in lines appeared without warning, chaos descended.[11]

Visibility on the ground presented a key issue for both armies. Soldiers struggled to adapt to the dark forest in daylight. Swan noted how, 'Excepting in the roads the dense wood rendered it impossible for any soldier to see what was going on three rods from where he stood.' In the absence of a visible enemy, imagination and fear ran roughshod over military discipline. For some troops, panic ensued. Plants became people and animals were mistaken for men about to attack. Nature mirrored the enemy, impersonated the foe. Schaff described 'dogwoods, with outspread, shelving branches, appearing at times through the billowing smoke like shrouded figures'. 'I wonder how many glazing eyes looked up into them and the blooming bushes and caught fair visions!' he mused. Struggling to make sense of the surroundings, soldiers fired on each other by accident. Confused by all the smoke, the 61st Virginia fired on the 12th Virginia. They hit James 'War Horse' Longstreet, their own commander. 'Of all the bullets in this wilderness doomsday volley the most fated was that which struck Longstreet,' told one soldier. Rescuers quickly, 'took him down from his horse and propped him against a pine tree'. Musket fire combined with forest fire, flames and smoke worsening visibility. One soldier described how 'in the breezeless woods, the powder smoke hung heavy, dense and low, arrested or clinging to the dwarf evergreens, which grew so thick that in many places they presented an impassable barrier to everything but the axe'. As regiments entered the Wilderness, they disappeared.[12]

As darkness fell, the doomsday landscape magnified. Farley wrote how, 'Night closed down upon an army enshrouded in a forest and oppressed with the sense of impending dangers.' Some troops sang to stave off the peril. Robertson reported the 'wild, weird music of that night' of 5 May. Schaff recounted songs sung at the Wilderness Tavern, where 'probably with feelings deeper than my own, the timber of the wilderness listened also'. The business of war continued in some reaches of the forest: the movement of ammunition and supplies, the discussion of tactics, the moulding of earthworks, and flash-fire engagements. On the night of 5 May, First Lieutenant Asa B. Isham's group took part in one skirmish. His experience was marked by incessant halts, musket sounds, falls into potholes and tripping over branches. Isham detailed a night 'monotony' broken only by falling over. Likewise the collecting of the dead proved treacherous. Rhea described how 'Rescue parties shrank from the dark woods for fear of rekindling the holocaust that had abated temporarily with the setting sun.'[13]

Other soldiers tried to sleep but were awoken by cries for help and musket fire. The doomed landscape haunted soldiers. First Lieutenant Oliver O. Howard recalled

> ... in the stillness of the night air every groan could be heard and the calls for water and entreaties to brothers or comrades by name to come & help them ... I will remember that at midnight, when I lay down to rest, and on waking during the night, their cries were ringing in my ears.

Darkness, fire, fear and death all mingled.

> Both armies lay exhausted amid forest fires that gave a final touch of horror to the stricken battlefield. The moaning of the unrecovered wounded, many of whom perished hideously in the spreading flames, mingled discordantly with the chorus of whippoorwills and other nocturnal creatures of the forest.

wrote Edward Steere. Nature and death came together.[14]

Only fire truly interrupted the Wilderness scene. Robertson told of how,

> gloomy woods in total darkness, except where lighted by the flames which belched from the muzzles of the thousands of muskets; the great sheets of fire, like flashes of summer lightning, lighting up the pall of sulpherous [sic] smoke which added to the dark gloom of the surroundings.

Asa B. Isham renamed fire the 'demon of destruction ... floundering and belching out tongues and volumes of flame in the murky depths below'. Fires spread in the woods. Inescapable flames cast the Wilderness as Dante's Inferno, an infernal blaze. Muskets fired. Artillery shells exploded, igniting the undergrowth. Log breast-works caught fire. Major General Hancock reported the failure of one defensive line due to 'a mass of flames which it was impossible at that time to subdue', just 'at the critical moment of the enemy's advance'. Dead leaves and spent shells stoked the blaze, the decay of the organic and the detritus of war escalating the conflict. Nature and machine alike fuelled the flames. Farley explained how fires colonised the buffer zones between the two armies, spreading

> unchecked over acres, disfiguring beyond possibility of recog-
> nition the bodies of the killed and proving fatal to hundreds of
> helpless wounded men who lay there looking for the friendly
> aid which never came and who died at last the victims of the
> relentless flames.

The injured struggled to escape. An estimated two hundred men were burnt alive on the night of 6 May. C. F. Atkinson described troops keeping cartridges back for their own use in case wounded or abandoned in the field.[15]

In attempts to capture key strategic points to gain victory, both armies manufactured intense doomsday landscapes. Union regiments attempted the taking of Saunders Field, a rare clearing in the Wilderness that once served as a corn patch for farmers. Rebels lurked in the woods awaiting the Union attack. Some of the worst fighting occurred at Saunders Field. The 140th New York Division crossed the field and received a vicious retort. One New Yorker told how, 'The regiment melted away like snow. Men disappeared as if the earth had swallowed them.' Fully exposed, Union men proved easy targets for Confederate snipers. One rebel related, with some remorse, that 'The enemy's ranks were as thick as blackbirds. Their flank was exposed to our brigade and the way we poured lead into them was a sin.' The situation worsened as Union artillery fired on Union regiments by mistake. Smoke from big and small guns alike settled in the air. Parts of the field caught fire, with the injured caught in the blaze. Cries of pain rang out across the grass. One regiment lost 268 out of 529 men in a single movement. A survivor recounted how 'The gloom which settled over us was unspeakable.'[16]

Wilderness and savagery

The Virginia Wilderness thus emerged as a doomed and hellish place for most soldiers. Warm climate, poor visibility, dense foliage and incessant flames left men distraught and tired. Old folklore also contributed to the catastrophe culture. Robertson despised the 'dreary and dismal woods', his regiment responding with great anxiety when told to 'push forward in battle array into the darkness and gloom of the thicket'. Soldiers recorded strange powers at work. Schaff elaborated on a 'mysterious silence' in the Wilderness, spoke of the ghost of legendary general Stonewall Jackson, and even felt that a 'spirit of the Wilderness is brooding'. The psychological trauma of conflict tied to imaginary forces in the wild. One soldier noted the 'nervous strain' before battle while hearing frightening noises in the forest. Colonel Thomas L. Livermore described a 'strange lethargy' taking over the Union generals, putting it down to the 'mysteries of the woodland'. Soldiers were trapped in a wartime culture of catastrophe. [17]

Union troops found it particularly distressing to fight on a site of former sacrifice where, just a year earlier, the campaign of Chancellorsville had dramatically faltered. Old knapsacks, scabbards and belts lurked in the undergrowth, decaying like leaves. Nature's spring beauty contrasted with the spoils of past conflict, one soldier expatiated how wild azaleas thinly disguised the 'debris of war'. Troop manoeuvres uncovered the archaeology of death. While wandering across the battlefield, Private George Tate found

> A skeleton lying by a large fallen tree in the woods . . . identified as H. Heyl, a member of Company H. of the regiment, known to have been mortally wounded the year before, by a cup, bearing his initials, found by his side.

Hancock's troops discovered 'The ground around their campfires, and for that matter everywhere, was strewn more or less with human bones and the skeletons of horses.' Soldiers spoke of finding over fifty skulls 'their foreheads doming in silence above the brown leaves that were gathering about them', a synthesis of dead nature and war articles. Campfires burned while 'half-open graves, displaying arms and legs with bits of paling and mildewed clothing still clinging to them' surrounded regiments.[18]

Scripture also guided perception. Scared at the idea of the Wilderness engulfing them, troops interpreted the battle by biblical reference. Robertson declared, 'Wild beasts and deadly serpents had their homes here, but none more fierce and deadly in their venom than the men in God's image who were rapidly moving into those dark defiles from both sides of the wilderness.' One officer told how his men advanced into 'Destruction's jaws into the devil's den'. The Wilderness resembled a biblical valley of death, a ground zero of mortification and sacrifice. Entranced by the 'gloomy region of death', Robertson went so far as to label the 'infernal' Wilderness 'indeed the very "Valley of the Shadow of Death"'. For some, this evil had purpose. Schaff posited that the deaths amounted to nature's revenge for slavery.

> I think I can hear the Wilderness exclaim with holy exultation, 'Deep as the horrors were, the battles that were fought in my gloom were made glorious by the principles at stake: and I cherish every drop of the gallant blood that was shed.'

he claimed. Schaff constructed a dark landscape of revenge and retribution. 'Dance on, repugnant and doomed creature! The inexorable eye of the Spirit of the Wilderness is on you,' he wrote.[19]

The scale of death and the horror of the battlefield shocked and dismayed all. One soldier related, 'The dead lay thickly strewn among the trees – the Wilderness was throbbing with the wounded'. Corporal Charles Smedley was horrified when the time came for his unit to advance: 'we started, going over the dead and wounded who had been slaughtered in the hundreds, and lay thick on the ground we went over'. Men lay slain in the killing fields. Overwhelmed by the 'place of slaughter and gloom', Robertson's 'regiment shivered with grief' on discovering the 'upturned ghastly face of their loved commander, with the life blood still oozing from the ghastly wound'. Horses were shot and maimed, Isham remembering how 'The neighings of the poor beasts, as they dropped by the wayside, were almost human in their plaintiveness.' For Robertson, mortality was everywhere: 'death flitted from bush to bush, and every thicket sheltered a corpse, while agonizing groans and cries of the wounded, were constantly ringing in our ears'. He told how 'Brave men were falling like autumn leaves, and death was holding high carnival in our ranks.' Dead soldiers and dead leaves decayed on the forest floor. Later photographs highlighted skulls resting among leaves, a union of dead landscape and dead people.[20]

Nobody seemed able to escape this world of savagery. One soldier described the yells of dying troops as a haunting 'concert of death'. William McDermott, chief surgeon for the Union 1st division, reported the treating of 139 Union and 69 Confederate injured, noting how 'a large number was found in the woods'. McDermott worried that a 'majority of cases were of such a nature as could not in my opinion have been moved in the hurry of the 7th'. Under fire, Isham sought cover beneath dead horses, and 'slept under the entrails of disemboweled steeds'. Isham declared his 'snug berth under the lee of a dead horse, giving off odors of putrefaction, is not to be despised when wild picket firing, at short range, is indulged in'. Under the onslaught of a Confederate battery 'which occasionally threw shells and solid shot in good range over us, cutting off the limbs of the trees', Smedley hid in a ditch, protecting his own limbs, until captured by rebel troops. Soldiers from both sides found new roles. Rather than bugling a Union victory, George the musician found himself nursing the sick. Employing his usual lyrical metaphors, George related how, 'our boys this morning were pretty near played out. Indeed I do not know what the wounded would do if it was not for the musicians.'[21]

Both sides emerged from the Wilderness at a loss. With a total of around 30,000 casualties, Schaff labelled the campaign a complete 'disaster'. Robertson called it 'the strangest and most indescribable battle in history. A battle which no man saw, and in which artillery was useless.' Grant moved on to fight Lee at Spotsylvania just days later, and a year later, the North won. With more men (twenty-two million to nine million) and industry (110,000 factories to 18,000), the energy landscape of the North proved far greater than that of the South.[22]

Reports emphasised the bravery of troops taking on the dark spirit of the Wilderness. Union General James Samuel Wadsworth was portrayed a hero in the wild, 'amid the smoke-filled thickets of the Wilderness, his spirit fired by the desperateness of the need, he led his men in charge after charge'. Henry Wing, journalist for the *New York Tribune*, fashioned the 'dreadful Battle of the Wilderness' into a nineteenth-century Bermuda Triangle of anxiety and loss. Wing wrote:

On May 4, 1864, a great army of citizen soldiers, comprising representatives of hundreds of thousands of families from every northern community, had vanished without warning, leaving absolutely no sign of their destination or hint even of the direction in which they had disappeared. There followed three or four days of such heart-breaking apprehension and bewilderment as the loyal nation had never before experienced.

Other commentators portrayed the battle with a sense of a historic, nationalistic and almost religious mission. General Horatio King granted the Battle of the Wilderness an epic dimension when describing, 'the scene of a contest the like of which has not occurred since Hermann destroyed the Roman legions in the forest of Teutoberg'. Schaff indulged in quasi-religious metaphor, portraying an evil spectre over the Union camp: 'face to face with disaster. What, what is the matter with the Army of the Potomac? Was an evil, dooming spirit cradled with it, which no righteous zeal or courage can appease?' The Virginia Wilderness served as a doomsday landscape, a place of death, disaster and slaughter brought on by a combination of difficult nature and brutal men. It provided a snapshot of the end of the world, an example of things to come. The Battle of the Wilderness showed the capability of American rank-and-file for wholesale destructiveness.[23]

Notes

1. Lisa M. Brady, 'The Wilderness of War: Nature and Strategy', *Environmental History* 10/3 (2005), p. 439.
2. Gordon Rhea, *The Battle of the Wilderness, May 5–6 1864* (Baton Rouge: Louisiana State University Press, 1994), p. 35; Ted Steinberg, *Down to Earth: Nature's Role in American History* (Oxford: Oxford University Press, 2002), pp. 90–2.
3. 'In Memoriam, James Samuel Wadsworth', (Albany: J. B. Lyon, 1916), p. 92, Huntington Library (herein 'H'), San Marino, Call No. 95437; Morris Schaff, *Battle of the Wilderness* (Boston: Hougton Mifflin, 1910) H 78739, p. 62.
4. Edward Steere, *The Wilderness Campaign* (Harrisburg: Stackpole, 1960), p. 1.
5. 'In Memoriam' (1916), p. 94; Schaff (1910), p. 107; Robert Stoddart Robertson, 'From the Wilderness', *The Ohio Commandery* (1884), p. 17.
6. Moses Kingsbury, 'Diary Jan. 11–June 11 1864', Huntington.
7. Ezra K. Parker, *Personal Narratives of Events in the War of the Rebellion* (1909), p. 14; Rhea (1994), pp. 29, 357.
8. Schaff (1910), p. 275; Porter Farley, *The 140th New York Volunteers* (1864), H 98134; Emil and Ruth Rosenblatt, eds, *Hard Marching: The Civil War Letters of Private Wilbur Fisk* (1992), p. 215; 'In Memoriam' (1916), p. 101.
9. Philip Racine, ed., *Gentlemen Merchants* (Knoxville: University of Tennessee Press,

2008), p. 647; The Military History Society of Massachusetts, *The Wilderness Campaign*, Papers 4 (Boston: 1905), p. 97; Rhea (1994), p. 297.

10. James Herbert George, Papers (Folder 48) 4–7 May 1864, Huntington; Rhea (1994), p. 125; Leigh Robinson, *The South Before and After the Battle of the Wilderness* (Richmond: James Goode, 1878), p. 39.

11. George, Papers; 'In Memoriam' (1906), p. 100.

12. Military History Society (1905), p. 145; Schaff (1910), pp. 59, 277; John Watts De Peyster's Notebook, *The New York Citizen*, 7 Jan. 1871, p. 347, Huntington; 'In Memoriam', (1916), p. 97.

13. Farley (1864), p. 17; Robertson (1884), p. 8; Schaff (1910) p. 99; Asa B. Isham, 'Through the Wilderness', *The Ohio Commandery* (1884) Vol. 1 H47977; Rhea (1994), pp. 250–1.

14. Military History Society (1905), p. 101; Edward Steere, *The Wilderness Campaign* (1960), p. 1.

15. Robertson (1884), p. 13; Isham (1884), p. 4; Major General Hancock, 'Report of Operations of the 2nd Army Corps' (1879), p. 13, Huntington; Farley (1864), p. 35; C. F. Atkinson, *Grant's Campaigns of 1864 & 1865* (London: Hugh Rees, 1908), pp. 195–6.

16. Rhea (1994), pp. 150, 152, 168.

17. Robertson (1884), p. 10; Schaff (1910), p. 87; Atkinson (1908), p. 200; 'In Memoriam' (1916), p. 95.

18. 'In Memoriam' (1916), p. 96; George Tate, 'Diaries & Military Records 1856–1908', Huntington; Schaff (1910), pp. 119–20.

19. Robertson (1884), pp. 10, 16–17; Watts notebook, p. 363; Schaff (1910), pp. 341, 274.

20. Watts notebook, p. 347; Corporal Charles Smedley, *Life in the Southern Prisons* (1865), p. 15, H 83016; Robertson (1884), p. 13; Isham (1884), p. 8; Robertson (1884), p. 12.

21. Rhea (1994), p. 183; William James McDermott, Letter to Surgeon A. N. Dougherty, 28 May 1864, McDermott Papers, Huntington; Isham (1884), p. 8; Smedley (1865), p. 15; George papers.

22. Schaff (1910), p. 275; Robertson (1884), p. 15.

23. 'In Memoriam' (1906), p. 121, p. 92; Henry E. Wing, *When Lincoln Kissed Me: A Story of the Wilderness Campaign* (New York: Eaton & Mains, 1913), p. 9; Schaff (1910), p. 326.

CHAPTER TWO

The End of the (old) World

Manufacturing the end of the world

The Battle of the Wilderness was not the only landscape of death. Nor was the Civil War the only battle raging. In the latter half of the nineteenth century, American business figures, industrialists and politicians dedicated themselves with fervour to nation building. Manifest destiny marked the day, God recognised as witness and guide to an expansionist mindset and process. The construction of a new American landscape of railroad connections, factories, and metropolises emerged as a high priority. A place of material abundance served as the ideal. The rise of this modern America transformed economy, society and environment. It also produced catastrophic consequences.

The construction of an urban and industrial America involved mass environmental disregard. Fur companies and hunters dedicated themselves to death to manufacture modernity. Manifest destiny translated in environmental terms as manifest doomsday. The manufacturing of a new world was built on the extinguishment of the old one. In the late nineteenth century, the American doomsday project gathered speed.

The costs of 'modern America' proved highly visible and yet, with few restrictions, spiralled out of control. Old-growth forests became timber. Engineers dammed rivers and harnessed water flows. Minerals were tapped and soil eroded. Hunters shot birds to extinction. Such enterprises entailed human as well as environmental costs: mining calamities included Granite Mountain, Butte where 168 died in 1906 and an underground explosion in 1907 at mine networks owned by Fairmont Coal Company in Monongah, West Virginia, that left 362 dead. The American enterprise hurtled towards environmental collapse. The railroad was one of the many doomsday machines servicing this effect. Criss-crossing the landscape, the steam locomotive linked economies of waste and destructiveness. Coal fuelled the machine, while refrigerated dead animals travelled on board. Gazing out of the windows of the train, tourists pointed their guns at wildlife. On the Great Plains, railroad rifles targeted the great American bison.

'The buffalo's doom'

'The buffalo's doom is sealed', wrote George Catlin, American painter and traveller, on an 1832 trip to the northern plains. Catlin twinned the survival of plains Indians with the life of the bison. With bison fur popular in the urban east and Indian wars spreading on the plains, Catlin fretted over the extinction of both. He famously volunteered as answer, 'A *nation's Park*, containing man and beast, in all the wild and freshness of their nature's beauty!' As fur trappers and traders travelled west, as settlement flourished, the fall of the bison seemed the more likely outcome.[1]

Steam-powered locomotives, mobile doomsday engines, expedited the killing process. Rail connected meat to the market, with bison wares leading to financial bounty in cities. Rail serviced the business of killing, the capitalist fortunes of modern Americans reliant on a new mass transport system. Advertisements broadcast widely the opportunity for slaughter. One poster promoted a 'Railway Excursion and Buffalo Hunt' from Leavenworth and Lawrence, bound for Sheridan:

> Hunt on the Plains: Buffaloes are so numerous along the road that they are shot from cars nearly every day. On our last excursion our party killed twenty buffaloes in a hunt of six hours. All passengers can have refreshments on the cars at reasonable prices.

A visit to Wild Wyoming promised excitement and adventure.[2]

Initially, hunting seemed to have little impact on animal numbers.

Figure 2.1 'The Far West – Shooting Buffalo on Line of the Kansas–Pacific Railroad', *Frank Leslie's Illustrated Newspaper* (1871) (Library of Congress).

An estimated eight million bison resided on the western plains in the late 1860s. The sheer number of bison even presented danger for rail passengers. In 1871, bison derailed two engines near Topeka, Kansas in just seven days. Reportedly, with 'herds running into and piling up before them . . . engineers learned to give a running herd "right o way"'. Such physical clashes between animal and machine symbolised a greater battle over the West. Within a decade, the tempo shifted, with the machine clearly ahead. The fall of the bison marked the rise of technological dominance. One postcard depicted tourists sticking their heads and guns out of train cars, taking potshots at the bison. The animals fell easily, without fight or gumption. An 1871 poster for Kansas–Pacific showed market hunters spraying bullets into black bison encircling a locomotive. The image paralleled popular portrayals of Custer's Last Stand (1876), whereby Indians encircled General George Armstrong Custer. Bison surrounded the doomsday engine and marksmen, threatening to engulf them – but, as in the case of the Battle of the Little Bighorn, the reality of the experience contradicted popular imagery: native peoples and animals were always the doomed parties. Across the West, the industrial locomotive with its armed passengers cut through bison ranks, charting a fast track to modernity. The railroad offered kill tallies, fun shots, and an entertaining ride. Easterners welcomed such invitations to doomsday, cheered on the bison collapse.[3]

Hunters, trappers and settlers already in the West equally enthused over the big hunt. The Bison Rush of the 1870s replaced the Gold Rush of the 1850s. An animal rush replaced a mineral rush. On just one 'Buffalo Harvest', Frank Mayer and Charles Roth accumulated 198 hides, each robe gaining them $3. 'They were walking gold pieces,' raved Mayer about the buffalo. Hunters classified animals in exclusively monetary terms, denying the living creatures status or ecological value. The mass excitement, or bison fever, of the period encouraged men to leave their jobs and families in the East to pursue the almighty bison dollar. Caught up in the growing storm, Mayer and Roth related 'how crazy the Western half of the country went over the buffalo rush'.[4]

The ensuing animal massacres connected the wilderness to the city, the demise of nature to the rise of the urban. Dodge City, Kansas, harboured a reputation for killing, of shoot-outs and duels, of 'a man dead before breakfast'. A mythic place of perpetual violence, the town was emblematic of the gun-toting Wild West. Dodge was certainly about death, but of animal, not man. Few cowboys were ever shot in Dodge but thousands of animal corpses made their way through the town. A hunter's paradise surrounded Dodge, a killing spree financed the town. One Topeka newspaper related how, 'Every ravine is full of hunters, and camp fires can be seen for miles in every direction. The hides of fourteen hundred buffalo were brought into town today.' Buffalo hunter Henry H. Raymond recorded in his 1872 diary a long list of kills. A typical day read: 'Went out, helped skin 17 buffaloes. Very warm and pleasant.' Entries detailed his shooting prowess, skinning times, weather changes, creatures (other than bison) spotted in the wilds, and how many Indians he saw in the vicinity. Raymond's kill tallies proved positively meagre compared to some: the local paper enthused how, in November of the same year, one Dodge man managed a cull of a hundred bison in a single day. Just outside of Dodge, Thomas C. Nixon slaughtered 120 bison in forty minutes. High kill tallies were deemed eminently newsworthy, and consistent reportage revealed widespread enthusiasm for slaughter. Journalists advertised and promoted the animal massacre. The modern repeating rifle, meanwhile, made such incredible kill tallies possible. Dubbed 'the Gun that Won the West', the 1870s Winchester rifle served as an efficient doomsday gun. The railway locomotive, the doomsday engine, carried the meat and skins to eastern markets. Between September and December of 1872, the railroad shipped 43,029 buffalo hides and 1,436,290 pounds of buffalo meat out of Dodge. A later Fred Harvey postcard depicted a huge pile of skins, with the caption, 'Forty Thousand Buffalo Hides Ready for Shipment. Dodge City, Kansas, 1878.' Men sat on top of totemic mounds of hides and bones, victors aloft their prizes. Pictures of mountains of bison skulls highlighted the success of the hunting endeavour and cast the animal as trophy, dollar and economic resource. The mountains also testified to the genocidal imperative, the doomsday at work.[5]

Warnings of animal extinction went unheeded. Along with Catlin, other travellers, scouts and hunters issued statements of concern. Plainsman Josiah Gregg wrote in 1845, how, 'The slaughter of these animals is frequently carried to an excess which shows the depravity of the human heart in very bold relief.' Hunters such as Buffalo Jones warned of imminent collapse in herds. Frank Mayer lamented how on his travels, 'all I saw was rotting red carcasses or bleaching white bones. We had killed the golden goose.' With some gravitas, *London Field* magazine warned in 1876, 'The disappearance . . . will be a scandal to civilization, and a subject for underlying shame and remorse to the children of the men who did nothing to stay the hand of the destroyer.' In marked contrast to Catlin's painting *Buffalo Hunt Surround* (1832) that depicted Sioux and bison vying for authority over the plains, the dominant modern image of the West was one of white power. The plains of the 1870s seemed very different from the 1830s: bones, skulls and billowy sagebrush replaced native grasses and wandering herds. Native animal presence was diluted, reduced to skulls and skeletons. Such new iconography of the West demarked white control but also denoted a dead landscape, a shadowy spectre. The killing grounds of the plains joined other battlefield landscapes such as the Virginia Wilderness.[6]

In a report for the Smithsonian, William Hornaday estimated just 50,000 bison remained on the northern plains in 1881. In 1882, with the industry of killing peaked, thousands of hides made their way on the Northern Pacific Railway from Montana and the Dakotas eastbound. By 1883, just 200 bison remained on the plains, under the protection of the US Army in Yellowstone National Park. Catlin's fear of doomsday on the plains had been realised. An estimated population of sixty million had been wiped out. From free-roaming animal to park museum object, the buffalo's life was effectively corralled. The animal relegated to the past, a curiosity of the mythic frontier. As one newspaper commented in 1903, 'It is a fact that the buffalo has passed into history.'[7]

A number of factors enabled the mass slaughter of bison in the late nineteenth century, and the victory of a Euro-American doomsday machine over western skies. On a broad level, the killing of the buffalo tied with the conquest of the West and the vanquishing of the Indian. Slaughter fitted within a neocolonial process of doomsday making that combined capital, economics, religious direction, technology and white triumphalism. The wipeout of the bison helped kill off the 'Old West' by the 1880s, while also preparing the groundwork for a new West to fill the void. The traditional environment was replaced with a new one, bison with cattle, Indians with settlers. The massacres reflected the dominant attitude to the American West, and to America as a whole: land and resources appeared infinite and free. Dreams of superabundance, however, were proved false with the collapse of the most numerous animal on the plains.

Buffalo Bill's Wild West show well illustrated this narrative. In the 1860s, Bill provided bison meat to Union Pacific Railroad workers, with legend telling of the eminent hunter dispatching 4,280 bison in just seventeen months. Bill also served as guide for buffalo hunters in the period. When he turned his life into theatre, Bill manufactured the stage West into a realm of violence: a landscape fashioned by the bullet, by the kill (and not by the plough, as historian Frederick Jackson Turner suggested). In one of the show's great spectacles, the 'Grand Buffalo Hunt', Bill re-enacted one of his classic hunts on the plains. It showed white warriors dispatch hordes of wild animals. Tellingly, cattle 'played' the bison role, replacing the wild with the domestic. The old landscape died before the rise of the new.

The bison hunts also drew on a 'fever' or craze in people for killing, an obsessive excitement over combining massacre with money-making. Mayer and Roth observed the sheer numbers of armed hunters on the plains, that 'With that many men after him, the buffalo didn't really have a chance, and just a few years were enough to decimate the herds.' Mayer reflected on his own role in the massacre: 'I'm often asked now what my feeling is toward myself that I helped wipe out a noble American animal by being a sort of juvenile delinquent with a high-power rifle.' The hunter felt the mass killing 'shameless, needless', but intriguingly an 'inevitable thing, an historical necessity', situating the kill within the broader triumph of white civilisation over native. For many hunters, there were basic incentives that kept the 'craze' alive. In the 'Wanton Destruction of Buffalo' (1872), Henry Worrall noted four fundamental reasons for the feverish killing: 'pleasure', 'tongues', 'excitement', and 'robes to get whisky'. Mayer sensed buffalo men were 'just a greedy lot'.[8]

Harpers forwarded other explanations for the bison cull. The magazine noted the failure of wildlife law and protection, claiming that with it, 'we would not have witnessed the wanton extermination of the buffalo'. *Harpers* also blamed careless hunting, the settlement process and the rise of the cattle industry. The magazine warned that other extinctions lay ahead, that the bison story was far from unique. Accordingly, 'doomed elk' were soon to be 'hunted to death', grizzly and moose 'becoming exterminated', and all the nation's 'big game diminishing with terrible rapidity'. The death of the bison appeared a harbinger of greater doom.[9]

Dancing on the edge of doomsday

In 1883, the year of the bison collapse, a number of Euro-American hunters refused to leave the Great Plains. The hunters expected the bison to return, for herds to wander the region once more. They were not the only ones. For many Indian nations, the demise of the bison represented the end of native life. As Sitting Bull told, 'A cold wind blew across the prairie when the last buffalo fell . . . a death wind for my people.' According to Francis Haines, 'The buffalo had been the central figure in the Indians' whole pattern of existence, and its disappearance left a spiritually disturbed people, socially disorganized, and lacking a meaningful pattern for a new way of life.' The loss of the animal left an economic, cultural and spiritual void. The key symbol (or totem) of environmental and cultural health on the Great Plains, the death of the buffalo symbolised total collapse, an indigene doomsday. Equally, any hope for recovery for Indian nations came to ride on the return of the animal to the Great Plains. Within five years of collapse, the Ghost Dance had begun, with chants of 'the buffalo are coming' reverberating across Indian reservations.[10]

In January 1889, spiritual leader Wovoka entered a trance at Pine Ridge Reservation, South Dakota, during a solar eclipse. A vision came to him from the Great Father, forwarding a picture of dramatic change. The Great Father instructed Wovoka to teach a five-day dance to all Indians. By practising the dance, nations would hasten the return of their peoples. Wovoka was a pacifist, with no intention of sparking armed conflict. He instructed the local Paiute peacefully to begin. News of the dance reached neighbouring nations. As one Paiute described, 'the dance spread like wildfire'. Myriad Indian nations appropriated the dance. The Lakota Sioux (most likely) added the concept of the 'bulletproof shirt' and interpreted the dance as a way to extinguish the whites. Practically the dance serviced pan-Indian identity and unity at a time of crisis. Only a few nations opposed the Ghost Dance, the Navajo rejecting it out of fear of bringing back the dead.[11]

The Ghost Dance emerged as a creative response to a despoiled landscape. The catastrophe culture introduced at Jamestown, and led by Euro-American pioneers, travelled westward in the 1800s. A spectre of death followed the frontier process, bones on trails not just marking the hardship of the arid environment but the greater narrative of environmental loss and destruction. Native plants surrendered to exotics, wild bison vanished and free land was fenced in. The landscape shifted. Dubbed 'ecological transformation' and 'cultural exchange' by some, the process proved far less neutral, with one side of the exchange broken. New replaced old. The original ecology of a nation was swept away. Devastation proved commonplace.

The process reached its climax with the 1890 census recording new levels of white settlement and the end of the frontier. By the late 1880s, Indian nations had undergone huge losses through war, disease, crop failure and starvation. The advent of the Ghost Dance corresponded with the end of a complete population crash, a demographic event of worldly proportions. Euro-American diseases, such as smallpox, measles and whooping cough, decimated Indian populations. With 50 to 90 per cent death rates, many nations withered or died. Those that survived faced the challenges of reservation life, confinement, and cultural death. At the Cheyenne River Agency on the plains, environmental conditions proved desperate. Reports noted the disappearance of game, menial rations, and the failure of agriculture in the arid soil and harsh climate. Most Indian nations experienced such hardship. Outside the reservation, Native Americans witnessed a changing landscape marked by wolves hunted to death, grizzlies gone, and the bison decimated. Indians sensed the death of the Great Plains. Sitting Bull spoke of his environmental fears, that, 'The sun will burn up everything. No crops can be raised in our land. And no rain will fall for many months to come.'[12]

The Ghost Dance operated as a cultural response to this unfolding environmental catastrophe. It engaged with an overwhelming narrative of enviro-cultural decay. In a few decades, plains Indians lost their way of life and homelands. The Ghost Dance articulated a social commentary on the taking of their ways. One study of the Pai nation underlined the dance as a way of tackling the series of life threats to the Pai at the time: disease, land loss, family bereavement, and a loss of power and authority. The song thus served as a poem of dissatisfaction, deprivation and discontent. It was a creative outcome of environmental damage and change, tethered to a doomsday landscape. The dance amounted to a mourning song on the enviro-cultural destruction of a nation, capturing the juncture of defeat, ruin and death. The Ghost Dance captured a moment in time of total collapse.

Wovoka described the process of the Dance to the Paiute:

> Soon now the earth will die. But Indians need not be afraid. It is the white man, not Indians, who should be afraid, for they will be wiped from the face of the earth by a mighty flood of mud and water. When the flood comes, the Indians will be saved. The earth will shake like a dancer's rattle. There will be thunder and smoke and great lightning, for the earth is old and must die.[13]

Caught within the ritual, each ghost dancer enacted a theatre of loss, a dance of death, followed by an invitation for revival. By dancing, Indians responded in a physical fashion to the unfolding doomsday landscape, the coming of a dead realm of animals and men. They witnessed the end. As performers, they enacted the unfolding collapse. They participated in the doomsday invitation, and welcomed the afterlife, or post-apocalypse. The Ghost Dance thus became a performance situated on the brink of large-scale change. Like bison funnelled by native hunters to the cliff edge, before falling to their death en masse, now Native Americans hurtled towards the same precipice, and with the same result. It was a dance on the edge of doomsday.

Watching the Ghost Dance play out, the culprit for inviting doomsday proved obvious. The dance targeted the Euro-American. Whites represented the pestilence on the plains. With their lives, culture and environment devastated, Indians witnessed first-hand the power of Euro-American culture to disrupt on an unbridled scale. With the Ghost Dance, Native Americans dramatised the concept of (the white) man making doomsday. The demise of the Old World bore testament that man could indeed manufacture his own Armageddon. Then, according to Indian folklore, man's new world would also fail. For Alvin M. Josephy, the dance was 'foretelling an apocalyptic end to white civilization'. Other Indian folk narratives situated Euro-Americans as pivotal to the end of the world, one story telling how 'cities will progress and then decay to the way of the lowest beings'. Such criticism corresponded with the rise of the American city. The World's Columbian Exposition at Chicago of 1893 celebrated the 'white city' of the future. For Native Americans, such foundations lacked strength. Indians served as early commentators on the emerging doomsday culture of white America.[14]

The dance represented an act of rebellion against the unfolding doomsday scenario. Unable to fight colonisation successfully, and with few means of opposition left, Indian nations employed the Ghost Dance as a final act of defiance against American aggression. The dance reflected a stark and poignant rejection of white cultural and environmental dominance, and a denouncement of the new American landscape of churches, farms, cattle, and schools. One Native American related: 'We want the earth to be as it is. Nothing should break up the surface of the earth. We will not have schools, nor churches, nor farms, nor white men's houses, nor their ways of living.' The Ghost Dance activated red power against white power. The bulletproof shirts of the Ghost Dance borrowed traditional shamanic powers of invulnerability. The dead would return, all mighty, and the white conquerors would fall. As anthropologists Henry Dobyns and Robert Euler described it, 'This millennial religion was, like others of its type, concerned with the redistribution of power'. The Ghost Dance represented a deep longing for the end of white cultural hegemony and destructiveness.[15]

The Ghost Dance worked for Native Americans on a number of levels. It provided cultural, psychological and environmental rejuvenation. Practically, it kept ritual alive. Indians 'learnt' the dance from 'knowing ones', part of an oral and physical tradition passed between people and nations. The dance drew on classic Indian mythology, and preserved the historic Circle Dance as performance. It also resembled other movements. In 1870, Pasinte Wodziwob taught a similar dance to the Northern Paiute, with Tavibo, the father of Wovoka, one disciple. Twenty years on, Wovoka's teachings captured a larger audience because of easier travel links and a greater scale of cultural destitution. In 1872, the Earth Lodge cult of northern California encouraged the establishment of subterranean dance houses for Indians to survive the forthcoming apocalypse – sort of pre-nuclear bunkers for Armageddon survival. The Dreamer Cult led by Smohalla along the Columbia river in Washington State meanwhile spread a message to return to the ways of the past. Smohalla's ritual symbolised a 'rejection of what they considered the abuse of "Mother Earth" for profit, in saving the earth for the return of their dead ancestors'.[16]

In psychological terms, the Ghost Dance facilitated the processing of group trauma. The song provided a physical release from stress and frustration. It proved a transformative and personal experience: an immersion in a narrative outside daily hardship and routine. The dance rekindled, albeit briefly, ceremonial life, at a time when reservations, schooling and white control sought to extinguish it. In its cultural component, Michael Carroll termed the Ghost Dance a revitalisation movement.[17]

The Ghost Dance provided many Indians with the dream of a land reborn. 'Rooted in hard times and devastation', the Ghost Dance offered hope. As Paul Bailey noted, 'To a whipped, broken, defrauded people – herded into barren reservations by the unfeeling and victorious white man – it promised one desperate and final hope.' The Ghost Dance spread the idea that a huge environmental catastrophe would punish Euro-Americans. The conquered natural world would again have agency and seek retribution. Nations believed that an earthquake, a landslide or a cyclone would swallow up the whites. Nature would strike back. Indians thus got something from the catastrophe: the promise of rebirth.[18]

Wovoka related, 'The earth is getting old and I will make it new for my chosen people.' Wovoka promised the rebuilding of a nation. As Shelley Anne Osterreich explained, 'This religion of the Ghost Dance promised that all the dead families and friends would return.' Resembling a post-apocalyptic zombie landscape, dead ancestors would rise again. The dance met the desperate wants of

Indians for a return to a pre-European America. A blanket cure for all ills of the past, it promised salvation to a broken people.[19]

An ecologically framed event, the Ghost Dance offered an early environmental protest narrative. In the Arapaho Ghost Dance, the opening song related, '*He*, my children, here is the pipe. Now I am going to holler on this earth. Everything is in motion.' Within the ritual, nature served a key role. James Mooney detailed how the Arapaho interacted with the environment during each performance, referencing the sky, surroundings, and mud. Dances regularly centred on a cedar tree, a totem of 'enduring strength and lasting life'. Some ceremonies included the presentation of gifts to the 'sacred tree'. For the Arapaho, ceremonial garments carried symbols of thunderbirds, turtles and the sun. Pawnee clothing featured eagle and crow feathers. For Pawnee ghost dancers, the crow was there to 'help people find what is lost – lost beliefs, lost ways of life'. Dancers 'shook the earth' by pounding their feet. Content covered food, land, seasons, and the rejection of Euro-American agriculture. The Ghost Dance fitted the image of the 'ecological Indian', who recognised land and ecology as a single related unit. It also carried ideas of Euro-American population collapse, city failure, and environmental doomsday.[20]

The dance highlighted the deep ties and connected destinies of Indians and land. For most nations, the danger posed by white conquerors to animals, earth, and people proved indistinguishable. All faced the white apocalypse, and none could escape it. This made the impact all the worse. Dobyns and Euler wrote how, 'Had the Pai not had an emotionally and psychologically deep attachment to and dependence on their lands, the trauma of Anglo-American impact might not have been so sharp'. Instead, the environmental trauma was shared. It also meant recovery affected all parties. As Dobyns and Euler imparted, 'the Pai placed the recovery of their land at the heart of their concept of the Ghost Dance movement'.[21]

Russell Thornton labelled the dance a 'demographic revitalization' project for Indians but, more than that, the song functioned as an environmental restoration narrative. It introduced a 'save the earth' concept. Robert Lowie described how 'the need for renovating the earth is an old and widespread American Indian conception'. With the world on the brink of disaster, the Ghost Dance emerged at a crucial hour, forwarding an environmental plea compacted into a traditional ritual of protest. The dance provoked images of a new game paradise, the return of animals from below. Wovoka's first vision was of a 'grand eagle . . . that carried me over a great hill', where 'broad and fertile lands stretched in every direction'. It forwarded the restoration not just of warriors but of plants, food and animals as well. On the Great Plains, the bison would return, and with it, the old ways of life. As Paul Bailey described, Native Americans saw that the earth would heal, that 'It would be lush and beautiful once more – with buffalo and game again in abundance'. As Bailey explained, by dancing, Indians hurried 'to remake the world, to return it again to its pristine and unsullied beauty'. As Wovoka foretold:

> when the flood has passed, the earth will come alive again – just as the sun died and was reborn. The land will be new and green with young grass. Elk and deer and antelope and even the vanished buffalo will return in vast numbers as they were before the white men came. And all Indians will be young again and free of the white man's sicknesses – even those of our people who have gone to the grave. It will be paradise on earth![22]

The creation of this paradise entailed the fall of white civilisation. As Dobyns and Euler related, 'One primary goal of the Ghost Dance movement was the removal of the stress of the land and resource loss by removing the Anglo-Americans responsible through supernatural means.' Native Americans clearly felt that Euro-Americans lacked an understanding of Mother Earth, and could not be trusted with its upkeep. As Wovoka explained, the Great Spirit 'told me that the earth was now *bad* and *worn out*; that we needed a new dwelling-place where the rascally whites could not disturb us'. The return of paradise, meant the absence of 'white man's manufacture', a place where no whites were 'permitted to live there'.[23]

Euro-Americans responded with fear to the advent of the Ghost Dance. Some mistook it for a war dance, and a prelude to military conflict. The 'bulletproof' shirts of the Lakota Sioux suggested a war-like people, ready to fight again. Rhetoric of open defiance proved a source of alarm. Commentators highlighted the cultish, 'heathen' elements of the dance, interpreting it as 'a barbaric return to savage dances' within nations. Some declared the Ghost Dance a condemnable bastardisation of Christian worship. A Quaker magazine lamented that the dance was 'eagerly embraced by many among, particularly those of the more ignorant class and those who were opposed to the habits of civilized life'. Major General O. O. Howard spoke only of 'The Messiah Craze'. Occasionally, commentators offered an environmental explanation for the dance. In his annual report, Major General Nelson A. Miles noted the scale of degradation on the plains, the 'distressed condition' of the Northern Cheyenne, the problems of food crises, crop failures and the loss of buffalo. An Indian agent for Miles suggested that the Indians exhibited 'no great fear of death', and ultimately felt that it was 'better [to] die fighting than to die a slow death of starvation'. An interview with Hollow-Horn-Bear on the Rosebud Reservation suggested that the dance was fundamentally about food. Hollow-Horn-Bear believed 'that the Ghost Dance, which is popular because it is a feast which hungry and starving Indians are attracted, and where they are fed, would cease if the people received sufficient rations to live upon'.[24]

Euro-American colonists also rejected the idea of their own responsibility for ushering in environmental and cultural collapse. They resented the fact that many Native Americans rejected white culture. The West was settled for good reason. Cities, industry, agriculture all prospered there. The only role for the traditional Indian was in wild west shows by Buffalo Bill, the only place for bison as exhibit animals penned in at Yellowstone National Park. Events of the nineteenth century were to be celebrated. Instead of hurtling towards doom, the frontier heralded a new era of prosperity.

White Americans acted on their fears. The Bureau of Indian Affairs imposed a ban on the dance but with little effect. Newspapers called for action. One Mohave County petition read, 'Music hath charms to sooth the savage beast [*sic*] . . . And it is said that the most effective is the whistle of a well-directed bullet.' Accustomed to conflict, the military interpreted the dance as another stage in Indian warfare. The army began an occupation-style response and infiltrated the movement. One order issued on Chief Big Foot, leader of the Minneconjou, read, 'If he fights, destroy him'. Meanwhile some Indians related that whites themselves were 'crazy with fear'.[25]

On 29 December 1890, at Lakota Pine Ridge, the Seventh Cavalry surrounded a Lakota Sioux camp near Wounded Knee creek. Soldiers massacred the Sioux men, women and children. One hundred and fifty-three died. People were butchered like animals. Like hunted bison, Indians lay in the snow, frozen and dead. The

Wounded Knee massacre was symptomatic of the period, of the new ability of whites to unleash huge acts of unbridled destruction. The doomsday massacre of Spotted Elk's group visually resembled the Battle of the Wilderness twenty-five years earlier. Wounded Knee marked a dead landscape with snow blizzards raging across dead bodies. It was another symbol of doomsday, another landscape of death.

Together, the Ghost Dance and Wounded Knee represented the end of Native America. The two events marked the end of the trail for wild nature and the finale of old America. For Black Elk, haunted by the images of 'the butchered women and children lying heaped and scattered all along the crooked gulch', Wounded Knee amounted to the death of a dream and of a people. As Elk lamented, 'the nation's hoop is broken and scattered. There is no center any longer, and the sacred tree is dead.' Old nature died at Wounded Knee. In 1890, the United States census recorded the official closure of the frontier but equally denoted a more significant end point, that of the Old World. In 1914, *Life* magazine published a poignant image of clouds shaped as buffalo, riding the sky, above a dead landscape of skulls, two Indians looking on. The picture was entitled 'Paradise Lost'.[26]

Black clouds

At 1 p.m. on Tuesday, 1 September 1914, Martha the passenger pigeon lay on the floor of her cage, dead, in Cincinnati Zoo. Martha was the only remaining passenger pigeon in the world. She died at the age of twenty-nine. Two male birds, her 'husbands', died four years earlier. Clearly, not just an Indian nations faced extinction. Edward Martin, game collector, commented on the fate of the pigeon, it's 'as if the earth had swallowed them'. The passenger pigeon represented another casualty of American progress, of the new stage in disaster.[27]

The passenger pigeon at first represented a creature of the 'New World' and an exciting scientific discovery. Around three to five billion birds populated North America on the eve of European colonisation. One colonial settler in Virginia recalled, 'There are wild pigeons in winter beyond number or imagination, myself have seen three or four hours together flocks in the air, so thick that even have they shadowed the sky from us.' In 1643, facing famine,

Figure 2.2 *A Pair of Passenger Pigeons*, watercolour by John James Audubon (1785–1851).

Plymouth colonists moaned of huge flocks of pigeons sweeping down on ripened corn and consuming 'a very great quantity of all sorts of English grain'. In 1648, John Winthrop saw the arrival of the birds as a great blessing, 'it being incredible what multitudes of them were killed daily'. Cotton Mather welcomed the 'mighty flocks' as a cooking ingredient. Pigeons were soon being killed in thousands, 200 caught in just two minutes by one colonist's trap.[28]

As well as its contribution to bird pie, the passenger pigeon drew admiration for its looks from naturalists. Naturalist, artist and explorer Mark Catesby drew the bird in 'Pigeon de Passage' (1871). It sat, somewhat awkwardly, on oak leaves, next to acorns, sporting a fancy pinky hue. In the 1820s, John James Audubon painted two passenger pigeons together. He captured the iridescent colours of the male bird, its blue plumage and bronze neck. The pigeon pair lovingly shared food across two branches. As always with Audubon, the birds were removed from their real-world environment. The background was white and plain, a specimen-type approach. The Audubon Society later employed a more sombre and prescient image of the pigeon in a promotional leaflet, depicting a sole male bird on a barren tree with a grey sky of birds fast approaching.

In autumn 1813, Audubon travelled from his Henderson home to Louisville, Kentucky. On route, Audubon witnessed a pigeon migration that, for three whole days, turned the sky black. 'The air was literally filled with pigeons; the light of noonday was obscured as by an eclipse,' he enthused. Audubon described a mass community of pigeons, their nesting 10 miles in length and 3 miles broad. Individual trees carried some hundred nests. Across the eastern United States, pigeons gathered in vast numbers in deciduous forests, feeding on acorns, chestnuts, beechnuts, seeds and berries (and occasional insects). Then, in spring and autumn, huge colonies of birds migrated, travelling at 60 miles an hour, and turning the sky black. One Wisconsin nesting site attracted an estimated 136 million birds, across an 850 square mile site. One commentator dubbed it, 'Nature's wonderland'. For others, huge bird flocks presented a darkish peril. 'I was suddenly struck with astonishment as a loud rushing roar, succeeded by instant darkness, I took for a tornado about to overwhelm the house and everything around in destruction,' described Alexander Wilson. The black skies, full of pigeons, showed the power of nature.[29]

On route to Louisville, Audubon was not the only one to marvel at the pigeon skies. He responded to the great show with awe, others responded with the gun. Shooters gathered en masse on riverbanks and in fields, taking potshots at the pigeons, eager for pigeon pie. The artist described the ensuing massacre, 'authors of all this devastation began their entry amongst the dead, the dying and the mangled'. An odour of death lingered.[30]

By the 1800s, the passenger pigeon serviced the economy of slaughter in the east. The passenger pigeon provided states with cheap meat for the poor and a decent economy for hunters. Professional 'pigeoners' sold their wares in cities. Birds afforded some protection for the market economics of farms, and a mild diversion for most sportsmen. Pigeons presented men with an opportunity for machismo, the breaking of flocks for individual kill an exercise in gun culture and masculinity. The pigeon provided ample opportunity for social bonding through hunting.

The bird attracted a wide range of killing techniques. Methods of death included baited traps and knocking chicks from nests with sticks. A pinch of the bird's neck by thumb and forefingers finished off the act. Hunters utilised bird decoys and stool pigeons, with threads through their eyelids, tied down to lure other birds. People burnt sulphur below roost trees (the fumes encouraging birds to fall to the ground) while others set trees alight, providing a quick route to roasted pigeon. Some hunters employed huge nets to capture migratory flocks. John C. French related the typical hunting scene, of how

> Flights of wild pigeons that almost obscured the sun are recalled by the display of a large net, a stool, hubs, baskets and other equipment used by the pigeoneers in the days when the netting of pigeons was a business of raising of squabs today.

The favoured technology of pigeon warfare, as with other kills of the time, was the trusty rifle, used to shoot down birds en masse. One hunter, firing a shotgun twice into a tree, managed the fall of seventy-one pigeons. Alexander Wilson described in detail the 'engines of destruction' used to dispatch the avian foe.[31]

Popular breeding and nesting grounds became massacre sites, fresh ground zeros of animal extinction. The transformation of avian birth sites into dead landscapes proved remarkable. At Sauk County, Wisconsin, in 1871, one onlooker described the gathering for pigeons, that 'a stranger would have thought it about war-time . . . [everyone] had a gun or wanted to borrow one'. At Petosky, Michigan, locals celebrated the return of flocks each season, of 'pigeon years' where the birds settled, with nests stretching 28 miles in one year, and numbers up to forty million. But the excitement was over the kill, not the living spectacle. For every pigeon, there also seemed a hunter. One commentator at Petosky wrote in the 1870s that, 'The pigeoners were everywhere. They swarmed hotels, post office, and about the streets.' Hunters totalled 50,000 birds a day. One pigeoner claimed 984 birds in twenty-four hours. In 1881, 500 men were netting 20,000 birds a piece at Petosky. Escapee birds were tracked down and eliminated. One newspaper described the hunters as a 'reckless, hard set of men'.[32]

Fiction and social commentary captured the mood of the time. In *The Pioneers* (1823) James Fenimore Cooper enthused how 'pigeons seemed a world of themselves', detailing one 'flock that the eye cannot see the end of . . . shadowing the field like a cloud'. He then delineated the turning point, the decisive moment, when 'death was hurled on the retreat of the affrighted birds'. One Currier and Ives print, entitled 'Pigeon Shooting', depicted a man with a gun hiding in a bush, with the enemy, the pigeon, lurking in the wilderness. *Leslies Illustrated* newspaper published a more realistic portrayal of bird warfare in 'Shooting pigeons in Iowa' (1867). The smoke of guns mixed with birds falling in vast numbers. Black clouds of death marked the skyline.[33]

Passenger pigeons died at a phenomenal rate. In 1878 alone, locomotives shipped 1.5 million in the summer, with an estimated one billion slaughtered. The pigeon lost forest habitat through the timber trade while food and nesting sites presented perilous kill zones. The rise of agriculture left pigeons with scant territory, one commentator claiming that the 'converting of forests to farmland would have eventually doomed the passenger pigeon'. Bird colonies adapted to farmland but faced rifles as they fed on grain. Dependent on large flocks for breeding and survival, the bird neared collapse. A lack of conservation impulse hastened the demise. State laws protecting nesting came in too late and went ignored. The major cause of pigeon death remained the hunt.[34]

The mass killing of passenger pigeons resembled the market hunting of bison. As Charles Bendire highlighted, both species went from vast numbers to zero in a very short time. Bird and animal alike once symbolised superabundance in American life. They also disproved it at their funeral. Both were tied to the same doomsday themes of technological power, human wastefulness and the priority of market economics. Along with a misled belief in a never-ending frontier, conservationist Aldo Leopold felt that the pigeon died thanks simply to the 'avarice and thoughtlessness of man'. William Mershon declared, 'The history of the buffalo is repeated in that of the wild pigeon, the extermination of which was inspired by the same motive: the greed of man and the pursuit of the almighty dollar.' For Mershon, the death of the passenger pigeon proved that 'The American people are wasteful'. The price of such extinction: as low as 50 cents a dozen. The death of the passenger pigeon testified to American power over nature. As one pigeoner, Martin, explained, man demonstrated complete control (and stewardship) over his environment, that 'not even if the birds should be sacrificed . . . for man is above the beasts, and the "beasts of the field and the birds of the air" are given unto him for his benefit and his profit'. In terms of the ability of American society to manufacture environmental doomsday, what better evidence is there than exterminating the most populous bird in the country, of facilitating its oblivion. Extinction demonstrated a show of power, and a broader ecological trajectory.[35]

On 24 March 1900, Press Clay Southworth shot the last wild passenger pigeon near Sargents in Pike County, Ohio. He was just fourteen years old. It seemed even at the earliest age, some Americans became trained killers. Fellow youngster Sullivan Cook, Ohio, told how he was trained in the kill, 'My father would say, "Get the gun and shoot at every pigeon you see"'. Clay's bird became a museum exhibit. By the twentieth century, only a few captive birds remained. For a while, some hunters expected the creature to return, that as with the bison, the pigeon had just gone away. The passenger pigeon instead entered the realms of zoo protection, Noah's Ark-type conservation, and given a last fleeting chance at survival. Fourteen years later, Martha died. Frozen in a 300-pound block of ice courtesy of the Cincinnati Ice Company, Martha became a museum exhibit, stuffed and put on show at the Smithsonian. She became a genuine doomsday article.[36]

The days of black clouds of passenger pigeons passed. The loss of the bison and passenger pigeon denoted a broader loss of wilderness. Their deaths helped fuel the rise of the American city and national greatness. The mummified Martha herself was surrounded by the black smog of the city, the new dominant landscape, and a realm controlled by humankind. Black clouds of pollution, dust and poison took the place of Martha and co. in the sky. Outside the city, mechanical agriculture transformed the countryside and, in a few years, produced huge black clouds of dust and debris, thanks to the Dust Bowl. The hunt of passenger pigeons was replaced with clay pigeon shoots, live birds replaced with artificial nature. And, on occasion, Martha, the taxidermy pigeon, still travelled, leaving the Smithsonian twice for first-class air flight to San Diego and Cincinnati.

Notes

1. George Catlin, *Manners, Customs and Conditions of the North American Indians* (Mineola: 1973 [1844]), p. 263. For the history of the American bison, see Andrew Isenberg, *The Destruction of the Bison: An Environmental History, 1750–1920* (Cambridge: Cambridge University Press, 2001 new ed.) and Dan Louie Flores, 'Bison Ecology and Bison Diplomacy: The Southern Plains from 1800 to 1850' *Journal of American History* 78/2 (1991), pp. 465–85.

2. E. Douglas Branch, *The Hunting of the Buffalo* (New York: Appleton, 1929), p. 129.
3. N. Hudson Moore, 'The Passing Buffalo', *The Four-Track News* (New York: April 1903), Autry Library, LA.
4. Frank Mayer and Charles Roth, *The Buffalo Harvest* (1958), pp. 15, 27.
5. David Dary, *The Buffalo Book: The Full Saga of the American Animal* (Chicago: Swallow, 1974), p. 97; Henry H. Raymond, 'Diary of a Dodge City Buffalo Hunter' ed., Joseph W. Snell, *Kansas Historical Quarterly* 31 (winter 1965), pp. 349–51.
6. Dary (1974), p. 122; Mayer (1958), pp. 84, 86; Branch (1929), pp. 145–6.
7. Dary (1974), p. 120; Moore (1903).
8. Mayer (1958), p. 27; Henry Worrall's drawing in W. E. Webb, *Buffalo Land* (1872).
9. Franklin Satterthwaite, 'The Western Outlook for Sportsmen', *Harpers* (May 1889), pp. 873, 875, 878.
10. Dary (1974), p. 93; Francis Haines, *The Buffalo* (New York: Thomas Crowell, 1970), p. 208.
11. David Humphreys Miller, *Ghost Dance* (New York: Duell, Sloane and Pearce, 1959), p. 28.
12. See Henry Dobyns and Robert Euler, *The Ghost Dance of 1889 Among the Pai Indians of Northwestern Arizona* (Prescott: Prescott College Press, 1967), pp. 2, 36–47; Major General Nelson A. Miles, *Annual Report* (Chicago: Department of the Missouri, 1891) p. 13; Miller (1959), p. 40.
13. Miller (1959), p. 27.
14. *The Way West* ep. 4 PBS, dir. Ric Burns (1995).
15. Major General Howard, *My Life and Experiences Among Our Hostile Indians* (Hartford: Worthington, 1907), p. 473; Dobyns (1967), p. viii.
16. Shelley Anne Osterreich, *The American Indian Ghost Dance, 1870 and 1890: an Annotated Bibliography* (New York: Greenwood, 1991), p. 29.
17. Carroll in Osterreich (1991), p. 48.
18. Barber in Osterreich (1991), p. 48; Paul Bailey, *Ghost Dance Messiah* (Los Angeles: Westernlore, 1970), p. 7.
19. *The Way West* (1995); Osterreich (1991), p. xi.
20. Frances Dewsmore, *Cheyenne and Arapaho Music* (LA; Southwest Museum, 1936), p. 66; Miller (1959), p. 29; Alexander Lesser, *The Pawnee Ghost Dance Hand Game: A Study of Cultural Change* (New York: Columbia University Press, 1933), p. 69.
21. Dobyns (1967), pp. 2, 18.
22. Russell Thornton, *We Shall Live Again: The 1870 & 1890 Ghost Dance Movements as Demographic Revitalization* (Cambridge: Cambridge University Press, 1986), p. xi; Robert Lowie in Osterreich (1991), p. 69; Bailey (1970), pp. 7–8; Miller (1959), pp. 27–8.
23. Dobyns (1967), p. 48; James P. Boyd, *Recent Indian Wars* (1891), p. 190.
24. Eileen Pollack, *Woman Walking Ahead: In Search of Catherine Weldon and Sitting Bull* (Albuquerque: University of New Mexico Press, 2002), p. 117; Howard (1907), p. 467; Miles (1891), pp. 3, 16, 12.
25. Dobyns (1967), p. 37; *The Way West* (1995); Pollack (2002), p. 116.
26. John Neihardt, *Black Elk Speaks* (1932), p. 218; 'Paradise Lost', *Life* 24 September 1914, p. 529, Autry.
27. Jennifer Price, *Flight Maps: Adventures with Nature in Modern America* (New York: Basic, 1999), p. 4.
28. Clive Ponting, 'A Green History of the World', in Fred White, *The Daily Reader* (Cincinnati: Writer's Digest, 2009), p. 142; Albert Hazen Wright, 'Other Early Records of the Passenger Pigeon', *The Auk* 28 (1911), p. 356; Arlie Schorger, 'The Unpublished Manuscripts by Cotton Mather on the Passenger Pigeon', *The Auk* 55/3 (1938).
29. Audubon and Wilson in William Butts Mershon, *The Passenger Pigeon* (New York: Outing, 1907), pp. 28, 84, 19.
30. Audubon in Mershon (1907), p. 35.

31. John C. French, *The Passenger Pigeon in Pennsylvania* (Altoona: Altoona Tribune Co., 1919), ch. 30; Wilson in Mershon (1907), p. 12.
32. Price (1999), p. 19; Mershon (1907), pp. 82, 87.
33. Cooper in Mershon (1907), p. 44; Price (1999), p. 2.
34. See Smithsonian, *The Passenger Pigeon* pamphlet (2001).
35. Bendire in Mershon (1907), p. 61; Leopold: Price (1999), p. 7; Mershon (1907), pp. xxi, 104.
36. From Sullivan Cook, in Merson (1907), p. 163.

CHAPTER THREE

The Armageddon Experiment: Doom Town USA

The twelfth of August 1953: in response to the US nuclear operation 'Ivy Mike' in the Pacific Ocean, the Soviet Union explodes its
own hydrogen bomb, nicknamed Joe 4. Charting the dangers of
the new atomic era, the *Bulletin of the Atomic Scientists* moves
its doomsday clock forward, to two minutes to midnight, the
closest ever to Armageddon. The editorial in the September issue
of the bulletin describes how 'the hands of the clock of doom have
moved again. Only a few more swings of the pendulum, and, from
Moscow to Chicago, atomic explosions will strike midnight for
Western Civilization.'[1]

In many ways, the atomic bomb was designed to avoid the strike
of midnight, to prevent the very fall of Western civilisation and
circumvent doomsday. Funded in secret by Presidents Roosevelt
and Truman, the US Manhattan Project (1942–45) ushered in the
world's first nuclear device, nicknamed 'The Gadget', exploded at
the Trinity test site, New Mexico, on 16 July 1945. The founding
remit of the Manhattan Project was to defeat Nazi Germany by
supplying a nuclear arsenal ahead of the Axis powers. Capable of
putting an end to world war, the nuclear bomb initially symbolised
a saviour of civilisation, not a doomsayer.

Just a few months after Ivy Mike and Joe 4, President Dwight
D. Eisenhower delivered his 'Atoms for Peace' speech before
the United Nations, establishing a framework for the peaceful
development of nuclear energy. Cogitating on the nuclear arms

race between the Soviet Union and the United States, Eisenhower mused:

> To pause there would be to confirm the hopeless finality of a belief that two atomic colossi are doomed malevolently to eye each other indefinitely across a trembling world, this greatest of destructive forces can be developed into a great boon, for the benefit of all mankind.

The president pledged on behalf of the United States, 'its determination to help solve the fearful atomic dilemma – to devote its entire heart and mind to find the way by which the miraculous inventiveness of man shall not be dedicated to his death, but consecrated to his life.' The lustrous atom promised white cities, agricultural abundance, electricity 'too cheap to meter' and a better environment for all. The United States would play host to this 'sunny side of the atom'.[2]

By 1953, however, the darker side of radioactive isotopes had already seeped into the national psyche. In the guise of mushroom clouds, radiation burns and testing in the Pacific, the atom augured a vision of a future American environment based around devastation. As the favoured jousting weapon of the Cold War, the atom bomb rarely dropped off the cultural or political radar. Questions remained over whether the Americans, Soviets, or humanity in general, could be trusted with such a powerful piece of technology. Atomic testing triggered deep-seated fears of environmental collapse. For Donald Worster, the beginning of the modern environmental era corresponded with the advent of the A-bomb. For the first time, humanity held the power to destroy its ecosystem in an instant. Radiation perfectly symbolised a new type of environmental danger symptomatic of the post-1945 era. It presented an invisible, deadly and artificial threat to all kinds of biota, and appeared capable of altering the fabric of life on a genetic level. Fears arose of total destruction and complete contamination. The nuclear landscape of America contributed to the formation of a modern culture of environmental fear and anxiety. The ultimate weapon produced the ultimate Armageddon landscape.[3]

Released in May 2008, *Indiana Jones and the Kingdom of the Crystal Skull* provided a late fourth instalment in the Hollywood blockbuster series. Directed by Steven Spielberg, and starring Harrison Ford, the film opened with a spectacular set piece at

Nevada Test Site in the 1950s. Betrayed by George 'Mac' McHale, and literally on the run from Soviet troops, Indiana Jones sought cover in a mannequin-filled ghost town. The town provided shelter from the Soviet menace but not from the home-grown threat of the atomic bomb. Sirens sounded, a nuclear test began, and Jones, the all-American hero-professor, escaped death by hiding inside a lead-lined refrigerator. Propelled into the air by the atomic shockwave, the kitchen appliance saved Jones who awkwardly clambered outside to greet a desert holocaust. Not renowned for their realism, the 'Indiana Jones' series had nonetheless captured a classic scene of Cold War mayhem and nostalgia. The film scene fused real and imagined disaster. A fabricated suburbia, nicknamed Doom Town, existed at Nevada Test Site in the 1950s and served as a real-life set piece for atmospheric atomic explosions. The film hinted at several environmental themes of the period: the Cold War and nuclear fear; the historic rendering of the West as wasteland; and the dangers of total destruction, as demonstrated in the nuking of suburbia (and all its occupants and appliances). This chapter explores the construction (and destruction) of Doom Town USA, the showpiece post-1945 doomsday landscape.

State-sponsored doomsday and the Cold War

Between 1951, the first year of US continental testing, and 1963, the year of the Limited Test Ban Treaty, a total of 126 atmospheric tests were carried out at Nevada Test Site (originally Nevada Proving Ground). Each test featured a series of military, scientific and civilian goals. Designated 'open', a small number of tests granted access to the site to an exclusive set of civilians and press reporters who watched the explosions from a hillside nicknamed 'Nob Hill'. Images were then broadcast to national media. The Doom Town moniker applied to the artificial suburbs created for two tests, Annie, part of the Upshot–Knothole series of 1953, and Apple 2, part of the Teapot series of 1955.[4]

Doom Towns 1 and 2 formed part of the civil defence components of Annie and Apple 2. Established in January 1951, the task of the Federal Civil Defense Administration (FCDA) was to promote Cold War preparedness on the home front. The race to construct a workable civil programme lagged significantly behind nuclear weapons production. In budgetary and political terms, civil defence projects always played second fiddle to the nuclear arms race. Throughout the 1950s, the FCDA constantly felt under pressure to pick up the pace of civil defence. With just a bare skeleton of programmes in place, but Armageddon on the horizon, agency officials sensed an urgent need to amass millions of volunteers, supplies and shelters 'fast'. The FCDA situated Doom Towns 1 and 2 clearly within this home offensive. The agency anticipated that the explosion of nuclear suburbia would increase public knowledge of the continental testing programme overnight, and generate mass involvement in civil defence activities. Americans needed to be immersed in catastrophe culture and be ready for doomsday.[5]

Originally hoping to orchestrate its own 'open' nuclear tests, the FCDA found itself 'fitting in' with the US military testing programme throughout the 1950s owing to budgetary restraint. Liaising with the Department of Defense and Atomic Energy Commission (AEC), the agency appended a number of 'civil defense indoctrination programs' to nuclear tests. These programmes had their own individual operation codes. Doom Town 1 came under FCDA Operation Doorstep, Doom Town 2 under FCDA Operation Cue. Hence, alongside military aims, that included perfecting ground warfare, testing protective gear and observing 'the psychological effects on troops participating in the exercises', nuclear tests encompassed a range of civil uses: from exploring the 'best goods' to survive atomic attack to planning emergency response work. Despite very different mandates, the FCDA was never at war with its federal allies. As one AEC report from the 1953 series stated, the presence of the agency at Nevada Test Site was widely regarded as 'no great inconvenience'. Neither were test villages a total novelty for the United States military. In 1943, the US Army recreated German and Japanese villages at Dugway Proving Ground, Utah, to test the effects of incendiary weapons. Projects such as Doom Town, however, deliberately nurtured public interest in nuclear testing: events that the military, for understandable reasons, tended to keep under tight control. While the FCDA wanted people to 'see' doomsday for themselves, the military worried over who exactly was coming to witness atomic fireworks in the desert.[6]

The testing of 'small town America' aligned with various other efforts of civil defence outreach in the 1950s. These ranged from an FCDA-sponsored reprint of a *Look* magazine article on nuclear disaster (with the shock title 'We're wide open for disaster'), to

Alert America (1951), a travelling exhibit detailing a simulated attack on American soil. The 'home front' concept had been briefly explored in Project East River, a university-written blueprint for 'emotional management' in the Cold War era, but Doom Town represented the apogee of 'nuclear danger on the doorstep', of atomic catastrophe culture. The two open tests rested on the dual mandate of FCDA civil defence strategy, firstly to foster genuine fear in John Q. Public, in this case by showing nuclear war as eminently 'real', and secondly, to convince Americans that survival was possible if trained and prepared, in other words, that the nuclear arsenal of Joe Stalin could be defeated. GI Joe needed his civilian counterparts. Doom Town emerged from this mindset of motivation through fear. It amounted to a real-time display of nuclear theatre to reinforce the dual message of the FCDA. For civil defence purposes, a 'successful doomsday landscape' at Doom Town translated into an efficient anxiety trigger, a galvaniser for public concern and dedication to the home front.[7]

Constructing Doom Town

The FCDA began constructing Doom Town 1 in early 1953. The first town was on the small side: just two houses of 'typical American framed dwelling' positioned 3,500 feet (1,067 m) and 7,500 feet (2,286 m) from ground zero. Freshly painted and fully furnished, the homes quickly attracted occupants. L. A. Darling Co., from Bronson, Michigan donated mannequins, while J. C. Penney clothed them. Local store manager Hillman Lee commented how, 'The outcome of this test is unpredictable, but the results of the evaluation may be a powerful factor in deciding fashion trends in years to come.' Journalist Bob Considine similarly enthused:

> All known varieties of American clothing, excepting a mink coat, have been placed on the mannequins, and it is possible that out of the evaluation tests may come civil defense warnings as to how to dress in these atomic times.

The well-dressed mannequins granted the test a theatrical element, a plastic realism. The utility of the dummies in gaining information on thermal effects gave way to wider interest in how the blast dirtied clothes and faces, and tossed limbs. In two other tests of

Figure 3.1 Operation Doorstep (1953) (courtesy of National Nuclear Security Administration/Nevada Site Office).

the 1953 series, pigs dressed in uniforms provided a similar role. Outside the houses were parked a range of cars. To avoid unnecessary costs, the FCDA appealed to manufacturers and garages to donate modern cars (only 1946–53 models) for nuclear testing. Justifying the auto-geddon, FCDA officials explained that they would learn how cars handled an atomic explosion and how future designs could be improved to limit nuclear effect. In response, fifty cars, most with mannequin drivers, were sequestered near the township. Agency officials promised to return the nuclear-tested motors for 'public display'.[8]

The name of Doom Town emerged as a favourite in military and public circles alike. Tentatively cast by some newspapers as 'Survival City', a darker moniker seemed better suited to a place destined for devastation. Doomsday-laden rhetoric filtered into press coverage, granting the nuclear landscape a fatalistic, nihilistic quality. The existence of Doom Town linked well with the doomsday clock, another device for measuring Armageddon, and its shift towards midnight, or world's end.[9]

At 5.20 a.m. on the morning of 17 March 1953, a nuclear detonation of 16 kilotons (kt) occurred at Nevada Test Site. Similar in size to the bombs dropped on Hiroshima and Nagasaki, the Annie Test was small compared to recent events at Bikini Atoll (and a dwarf in comparison to the Soviet Union's Tsar Bomba of 1961 that measured 50,000 kt). Nonetheless, the nuclear exercise proved an ambitious affair. Fifty-five tanks moved towards ground zero as part of a nuclear battlefield manoeuvre code-named Camp Desert Rock V. Completing the scene were dogs held in harnesses as part of a radiation experiment, live sheep driving, bomb-shelter testing and the destruction of Doom Town. Nearly a thousand observers amassed at the News Nob 7½ miles from ground zero. FCDA administrator Val Peterson was among the atomic voyeurs. National television coverage spread the message of doomsday. Radioactive fallout swept across south-west Utah. As one observer remarked, 'it's a big fraternity, this order of the mushroom and its growing all the time'.[10]

This 'order of the mushroom' boasted a huge number of press officers. Three hundred covered the story from News Nob, including a young Walter Cronkite. Twenty sat with troops in the trenches, hearing Captain Harold Kinne brief soldiers, 'This is the greatest show on earth. You won't be hurt. Relax and enjoy it.' With a 'ringside' view, Robert Bennyhoff for the *Las Vegas Sun* pondered, 'How does it feel to get ready to sit on the doorstep of destruction?' and took solace in his first-aid kit and shovel. Excitement came from the expansive and innovatory television coverage established for the test, with three individual segments devoted to before, during and after the bomb. The *Las Vegas Review-Journal* highlighted how 'television audiences all over the nation will be given their first look at the havoc wrought by an atomic bomb'. Press agents negotiated with the AEC to linger in the blast zone. The commission later received numerous complaints over the awkward position of the dust cloud after the shot went off, blanketing Doom Town and preventing good camera shots. An official Operation Doorstep documentary followed. Billed as a 'shocker' by the press, the documentary corresponded with a national campaign for blood donation. Clips of the test were even employed in a three-dimensional film, entitled, *Doom Town: Death of a City by Atomic Destruction*.[11]

Doom Town 2 was largely a repeat of Doom Town 1, only on a bigger scale. The Apple 2 test incorporated a total of seventy-five military, scientific and civilian experiments, most prominently nuclear battlefield training and weapons development. Broadly stated, like before, the civilian aims were 'to assess the effects of nuclear detonations on civilian populations and to evaluate CD emergency preparedness plans'. A reporter covering the preparations described how 'Operation Cue has many phases, military, civilian and dietary' (the dietary aspect referring to tests on table food). To accumulate useful information on nuclear attack, experiments included the monitoring of animals in nuclear shelters, food in tins, and the construction of another Doom Town.[12]

Doom Town 2 featured ten buildings as compared to two. Along with a filling station, radio station, pylons and driveways, the second town far better resembled a typical suburban landscape of the 1950s. The main street, signed 'Doomsday Drive', gave a sense of order and place. The *Los Angeles Times* printed a map of the 'Mock City', while CBS filmed a city tour prior to the blast. Like before, each house featured furniture, food (including beer), appliances, motor cars, and even children's toys. One hundred and fifty industries contributed to the project. J. C. Penney from Las Vegas again clothed the residents, this time seventy-five of them, with $2,500 worth of clothing. Store manager Hillman Lee boasted how 'no two mannequins will be dressed alike', the dummies serving as proud icons of American individualism and consumer choice. The FCDA noted that the textile industry would be interested to learn the effects of fallout on clothing materials. Eighteen mobile homes dotted the landscape, furnishing an atomic caravan park adjunct to the central suburban hub. The proliferation of cars and caravans reflected the popularity of mobility in the period. The AEC billed Doom Town 2 as 'representing a typical American community' – a community about to go through Armageddon first hand.[13]

Interest in the open test peaked well before the dust clouds rose. Some 5,200 onlookers were expected on site, representing forty-two states of the nation. Fifteen hundred news reporters arrived and flooded nearby Las Vegas. Both CBS and NBC planned prime-time shows, including a *Today* show live from the nuclear trenches. The *Albuquerque Journal* predicted that, 'It will be the "most-covered" atomic test in history'.[14]

Like excited boys playing toy soldiers, reporters enthused over the novelty, scale and drama of the test. The article 'Newsmen join GIs in Shadow of Atomic Tower', proved reminiscent of wartime reportage with its references to tanks and military manoeuvres.

Figure 3.2 *left* Nuclear family (before). Operation Doorstep (1953) (courtesy of National Nuclear Security Administration/ Nevada Site Office).

Figure 3.3 *above* Nuclear family (after). Operation Doorstep (1953) (courtesy of National Nuclear Security Administration/ Nevada Site Office).

One reporter stationed with the military task force provided coverage 'from the frontline' of troops 'ready for nuclear warfare when it comes'. Journalists devoted little coverage to the practical dangers of radiation or questions of fallout. Instead, all kinds of science fiction superlatives and personality stories were thrown at Apple 2. Like a genie in a bottle, 'A nuclear weapon to be hurled Tuesday against a deserted American city on the desert northwest of here is expected to unlock further secrets of civil defense and point the way to survival in the atomic age' promised the journalist Archie Teague. 'It's "Out of This World" at Atom City', wrote one reporter on his first visit to Doom Town, describing a nuclear cinematic, 'civilization far, far away'. The atomic test appeared destined to bring America something special. At the very least, journalist John F. Cahlan expected a visual feast to rival the biggest Hollywood film, 'The world's most expensive premier will be unfolded out on the Nevada desert . . . and nothing Hollywood has ever produced will be able to equal it'. For Cahlan, Doom Town 2 was 'expected to be one of the greatest horror programs ever produced'. Marvin Miles from the *Los Angeles Times* cast the bomb itself as a technological villain, for 'it casts an aura of malediction across the vast sweep of the test range'. Journalists forged a sense of expectation around the experiment, portraying the bomb as the prime agent in a spectacular unfolding drama.[15]

High-speed winds and associated fears of drifting fallout delayed the Apple 2 test for ten days. The delays infuriated members of the press who struggled with what to write, where to stay and how to meet deadlines. More than two-thirds left. Those who remained fashioned the constant postponements into the real disaster story of Apple 2. Newspapers described 'mentally and physically exhausted' reporters making 'growls' and looking 'impatient'. Like some sort of wild animal hiding in the zoo, the atom failed miserably to entertain its guests. *Quill* magazine explained that the atom bomb had simply 'muffed a cue'. Las Vegas officials faced accommodation woes as the nuclear test ran into time allotted to a golf championship, a prizefight and the annual community parade, whereby Doom Town vied with the Vegas Helldorado as entertainment. The Sands Hotel cancelled bookings, explaining to its disappointed guests that its 'first responsibility at this time is to the National Defense effort'. United Press released pictures of Sands' Copagirl Linda Lawson by the poolside wearing an A-Bomb crown, sporting the title 'Mis-Cue'.[16]

On 5 May at 5.10 a.m., the 29-kiloton tower shot finally went off. A military task force (codenamed RAZOR), consisting of 1,000 troops, eighty-nine armoured vehicles and nineteen helicopters led an assault on ground zero, some tanks approaching to just 890 metres from the centre of the blast. Around 2,000 people watched Operation RAZOR from Mine Mountain where a public address system commented on the battles below. Civil defence workers entered Doom Town for radiation samples, the practising of first aid, and disaster training. The AEC dubbed the operation 'the most comprehensive civil-defense exercise held in Nevada to date'. Emergency food was flown in from San Francisco and Chicago. Radioactive fallout travelled outwards from the blast area towards St George, Utah, and beyond. Civil defence worker Miss Helen Leininger, from New York, positioned in the trenches, responded to the Apple plume with 'Gee, it's a beautiful thing'. Along with other observers, she took home an official Open Shot badge as a memento. It depicted a white mushroom cloud.[17]

Fear of fear

Doom Towns 1 and 2 signified potent experiments in nuclear fear and catastrophe culture. One planner pithily summed up the work of the FCDA as promoting 'bomb consciousness.' For Laura McEnaney, 'the FCDA functioned as an ad agency whose client was the bomb'. The selling of the atom proved a difficult business, often more about the production and management of fear than the advertising of a decent nuclear product. Seeking a fine, perhaps elusive, balance of push and pull, FCDA propaganda aimed to shock people out of post-war malaise and apathy while generating a 'healthy' level of fear and anxiety. The Doom Town projects fitted perfectly within this mandate.[18]

While keen to inculcate concern in its citizenry, the FCDA proved particularly wary over atomic fear escalating into mass panic. The spectre of *The War of the Worlds* radio broadcast of 1938 lingered: could Americans be trusted to respond in the right way to any serious threat? At times, officials seemed to fear fear more than the bomb itself. In 1953, the same year as the Annie Test and Doom Town 1, Val Peterson produced a pamphlet entitled 'Panic: The Ultimate Weapon,' printed in *Colliers* magazine. Peterson related how 'like the A-bomb, panic is fissionable. It can produce a

chain reaction more deeply destructive than any explosive known.'
For some officials, panic embodied a greater peril than a nuclear-
triggered end of the world. Predictions of chaos and pillage, of a
Hobbesian savage world, arose first in government and military
quarters in the 1950s, long before being picked up by Hollywood
film directors and their own visions of *Mad Max*-style barbarity.
Soviet nuclear warheads were viewed as psychological weapons
capable of striking fear into the heart of everyday Americans. If
nuclear fear grew too much, American mettle could shatter, with
the Cold War lost in an instant. Mob rule in the aftermath of disaster, a world of complete chaos and breakdown, proved the greatest
concern. Such horrors never materialised under the spectre of the
bomb but anxiety over social breakdown continued past the Cold
War, evidenced in the media portrayal of social and environmental
collapse with Hurricane Katrina in 2005.[19]

The US military took the psychology of nuclear fear seriously in
the 1950s, with biomedical experiments designed to observe troop
response and an 'indoctrination' programme calculated to reduce
anxieties. In a September 1951 report, officials spoke of developing
'psychological resistance to fear and panic' by developing an 'emotional vaccination'. Troops needed to hold firm before the apocalypse, and were deliberately placed close to ground zero during
the decade to 'evoke fear response'. To a degree, the Doom Town
projects served as a civilian equivalent of this 'fear evocation'.[20]

The FCDA approach to the management of catastrophe culture
also drew on wartime experiences. Practically, World War II
had established some acceptance for moral persuasion, with
the Psychological Warfare Division (sykewarriors) engineering
schemes to nurture public belief in victory and undermine German
steadfastness by propaganda drops. FCDA publications aimed to
reignite a degree of wartime anxiety, and with it, Cold War preparedness. Nuclear war thus compared with world war, situated
as something familiar, able to be tackled, and conventional. Hence
the 1950 pamphlet *Survival Under Atomic Attack* claimed, 'You
can live through an atom bomb raid and you won't have to have
a Geiger counter, protective clothing, or special training in order
to do it'. The use of the term 'bomb raid' conjured newspaper
headlines of European blitz-type attacks. The booklet normalised
the nuclear threat by stressing 'atom-splitting is just another way
of causing an explosion'. The fate of Doom Towns 1 and 2 thus
resembled the wartime fire bombing of Dresden and Coventry.[21]

The FCDA ultimately wanted a public ready to take on the nuclear threat, to 'be prepared' for attack. Expected to have faith in the defensive might of the United States government and military, Americans were also encouraged to take individual responsibility for their survival. This partly reflected the harsh reality of FCDA budget constraints (some 90 per cent less than requested from Congress) but also tapped a broader history of American dual responsibility. A protective governmental shield went hand in hand with staunch individualism. Within the nuclear propaganda of the 1950s rested a survivalist message coupled with a strong desire to propagate 'American will'. American citizens needed to toughen up for the Cold War. As Guy Oakes related, it was hoped that 'The discipline of civil defense would cultivate the toughness needed to meet the demands of the Cold War'. This sense of toughness translated into personal responsibility, community organisation, and trust in government, military and self. It also served the psychological effect of involving citizens in a Cold War marked by geographical remoteness.[22]

The Doom Town project marked a shift for the FCDA from simple to sophisticated tools of persuasion and fear generation. It showed the agency pushing into the mainstream of modern media. Pamphlets from the early 1950s, such as *Atomic Attack: How to Protect Yourself*, offered naive and simplistic takes on nuclear conflict. Gaining mass exposure proved initially challenging. A couple of years into the programme and the FCDA had moved up a gear. One million references to civil defence were found in magazines and newspapers for 1953 alone. The FCDA supplied the *Operation Doorstep* documentary to television stations across the nation, even making a small profit ($15,000) on the print. Over sixty radio programmes covered the 1953 Annie test. Doom Town was a far more modern and media-savvy product to behold.[23]

The Doom Town experiment added something new to the civil defence arsenal of the FCDA. For the first time, an American landscape had been destroyed by the bomb and witnessed by the whole nation. Doom Town provided a native, more poignant apocalyptic image to that offered by photographs of the wartime Japanese cities of Hiroshima and Nagasaki devastated by 'Model T' atom bombs. Press coverage frequently compared the two Doom Town tests to the two Japanese bombings, noting the similar size of the kiloton explosions. To a degree, Doom Town highlighted the potential for a nuclear-style Pearl Harbor, a Cold War surprise hit on the American mainland. The threatened townscape was something everyone could relate to. It was a quintessentially American environmental doomsday. It was on home soil. Doom Town proved a powerful, exceptional and successful doomscape. Such power rested not just on nuclear fear but tapped a historic sense of the West as wasteland, a contemporary fascination with suburbia, and nascent concerns over environmental collapse, themes that will now be explored.[24]

West as atomic wasteland

Prior to the establishment of Nevada Test Site United States atomic testing had proved a remote experience, something situated either as unknown and secret (as with the Trinity test) or 'comfortably away from home' in the form of military excursions to Japan and the Pacific. A mesmerising manifestation in the open sky, the atomic mushroom cloud had always lacked any strong landscape association at the base of the plume. On one level, transplanting testing to Nevada sought to continue this sense of anonymity. Scott Kirsch argued that AEC propaganda was 'designed, quite literally, to take the *place* out of the *landscape*', severing any ties between the bomb and human or ecological impact. As Tom Vanderbilt theorised, the test site signified a 'blank space' on several levels, blank on the map because of secrecy clauses but also cast as a blank canvas to shape (and bomb incessantly) by the military. Any sense of a regional past, in the form of Indian, ranching and mining heritage, was sacrificed to the overriding military function of the site. Already tied to the Las Vegas (later Nellis) Bombing and Gunnery Range in wartime, the atomic project reasserted the permanence of military control. Covering the 1953 Annie test, the *Wyoming Eagle* commented upon the destructive power of the atom, noting, 'It is a fearful thing to know that man has the power to destroy instantly thousands of persons, buildings, and belongings, leaving no trace of their existence'. But any meaningful pre-atomic existence had already been taken from the desert region. The NTS had no identity other than the one given it by the AEC and United States military. This geographic and historical cleansing allowed officials to cast the NTS as the perfect experimental landscape – neutral, uncomplicated, a desert Petri dish. Along with enabling military 'training and indoctrination', the AEC considered the Nevada Test

Site to be 'a backyard laboratory', a realm ripe for testing. It was a secret atomic landscape.[25]

And yet, explosions in the Nevada desert brought the bomb home. The annual test series connected the atomic age with frontier history and western pioneering, granting it a decidedly American identity. The military patina crafted on to the region was itself rich in frontier motifs and ideology. Nevada Test Site, and specifically Doom Town, gave the mushroom cloud a base, a desert garden to prosper in, and a realm of belonging.

This belonging drew on a West of frontier pioneering, escape, experimentation and knowledge accumulation but, equally, conquest, marginalisation, militarisation, and devastation. Testing activities called on the West of old and of present day. The power of the atomic doomsday image rested on themes that had shaped the broader western locale over several centuries. Practically, testing occurred in a vast military landscape, 1,350 square miles given over to weapons proving. The adjoining Nellis Air Force Range added more than 3,000 square miles to the equation. The vast militarised landscape signified one huge homestead. According to western folklore, only those desperately searching for escape and not wanting to be seen chose the most inhospitable places of the West to settle in. Hence, deserts attracted those looking for hiding places, such as the Mormons at Salt Lake in the 1840s or the military in Nevada in the 1950s. Equally, the use of Nevada Test Site signified the Cold War-inspired, gun-toting arm of the federal west. Washington bureaucrats and military generals exercised control over such territories. The continual bombing of the West reflected the power, might and disregard that Washington exhibited towards the frontier in the mid-twentieth century.[26]

The choice of Nevada for testing reflected a histrionic and enduring Euro-American culture of casting the American West as arid and unfertile. Deeply embedded within the 'doomsday message' of the US atomic bomb was the vision of the West itself as wasteland. Vanderbilt talked of how 'the country's most primitive terrain was suddenly lost to its most sophisticated technology' but the terrain had already been 'lost' long before the bomb. As author Wallace Stegner saw it, the West was 'the land nobody wanted'. It was not a place of survival but one sated in death and doom. Popular frontier folklore of the nineteenth century told of William Manley chewing bullets on the trail, the Donner Party collapsing in the Sierra Nevada. Paintings by Charlie Russell depicted the classic West as a landscape of sun-scorched skulls and cattle bones. The deserts of Nevada had very little to offer in terms of agriculture, industry, nature or survival. Why atomic testing failed to shock was partially due to the sense of darkness already hovering over the desert landscape – of a realm already spoiled and contaminated prior to the advent of the bomb. It was almost as if the region had already been nuked, with the net result of atomic pioneering being continuity, not change. To paraphrase native scholar Valerie Kuletz, Nevada Test Site represented a tainted landscape, and thus ideally suited to bombing. This facilitated the sense of the NTS as the perfect national sacrifice zone: worthless lands primed for Armageddon.[27]

The nuclear experiment also drew on a historic association of the West with frontier energies. Atomic pioneering in the desert signified the latest stage in developing the region. The nuclear project followed on from past trapping, mining, logging and drilling activities. The atomic landscape seemed strangely familiar for it replicated patterns of behaviour typical of frontier mining. The uranium boom of the 1950s provided a second mineral rush for the West, forging temporary townscapes like Moab, Utah, that sported single-industry focus and fervent business excitement. The *Denver Post* described Jeffret City as an 'atomic age frontier town', while scholar Michael Amundson nicknamed such places the 'yellowcake towns' of a new atomic frontier. With the promise of 'electricity too cheap to meter', hope arose of instant riches, not just for nuclear corporations but for the broader populace as well. As with the previous century's mining experience, the boom quickly turned to bust. By the 1980s, former uranium towns, such as Grants, Arizona, were struggling to survive from a marginal tourist trade. Ultimately, the quickest place of turnover was Doom Town itself, which literally (and visually) went from boom to bust in twenty-four hours. With a sense of romantic fatalism, one journalist described Doom Town as 'the city that was built to live for but one day'. Posting collages of live guests alongside dead mannequins, the *Los Angeles Examiner* read, 'A-Boom Town is a Doomtown', announcing the rise and fall of the atom in one pithy headline. Some journalists talked of the 'ghost community' gathered at the fake town site and experiences on the 'weary mule train' of the press convoy travelling there. Bikini woman 'brandishing a Geiger Counter as she checked the beard of a grizzled prospector for radioactivity' read one article. For the *Las Vegas Sun*, 'A modern day doom town in the Nevada desert' signified the classic story of pioneer boom and bust. Doom Town became a twentieth-century ghost town, an atomic twin to the deserted towns of fifty years before, a modern-day Rhyolite. It joined the ramshackle collection of other deserted frontier towns, gradually being worn away and retaken by sunshine and weeds. As the *Las Vegas Review-Journal* noted, 'Vagrant desert sands silted through the wrecked doors and shattered windows of Doom Town today, but the housewives made of this model city made no moves for their brooms.'[28]

The interface between old and new West played out spectacularly at the atomic desert. Paraphernalia of the nuclear age replaced the decrepit mining machinery of a previous century, new gadgetry replacing rusted technology. Site of the assembly of the first atomic bomb, the 1913 McDonald ranch house, next to Trinity, perfectly demonstrated the synergy between old and new West, by bringing wooden frames and Geiger counters together. The Atomic Energy Commission cemented the ties between the nuclear bomb and the folkloric West. A number of official publications included references to cowboys and guns while test site workers and scientists were lofted as the new hardy frontier pioneers. Be it Willamette Valley, Fort Sutter, or the barren lands of Nevada, land was there to be overcome, to be tamed, to be conquered. Many frontier projects entertained the prospect of turning the desert into something constructive. What made the bomb different was its capacity to destroy as well. A 'nuked landscape' was created, a novel geography crafted of the West based around doom.[29]

Making the desert boom

The desert scene of Nevada also suited the desolate imagery associated with Armageddon. The end of the world spelt nothingness and the absence of life. A historic landscape of 'emptiness' with clear cultural associations of death and decay seemed appropriate to Armageddon testing. With its aridity and sparse foliage, with its stories of hardship and dying, the desert appeared dead before nuclear destruction. Images of natural and nuclear desolation combined in complementary ways, provoking a degree of environmental synergy. As the *Los Angeles Times* reported on Shot Annie, 'the shot tower draws the eyes like an evil magnet, spiring pencil-like 500 feet above the dusty, desolate floor of this mountain-rimmed wasteland'. For the *Washington Post*, the doomsday landscape boasted 'an air of gross unreality', an otherworldly quality. Arguably, bombing a worthless military wasteland had little impact or meaning. Total destruction failed to matter or change anything.[30]

The tests equally amounted to the exercising of control over the natural world. On the surface, control equated to abject power, a wholly destructive display of military might whereby the environment fell victim to contamination. With its craters and bomb detritus, the 'proving ground' of Nevada served as visual evidence of this process. The atom displayed perhaps the ultimate manifestation of human and technological power over the natural world. Deep down, the relationship between atomic energy and the environment was far more complicated. Alongside the Nevada tests, the United States government began experimenting with civilian schemes that promised better agriculture and economic productivity through atomic engineering. A 1949 report by the AEC for the United States government detailed, 'how it opens new vistas of better health, of more abundant food supply, of greater control by man over his environment: and how – along with its promised benefits – it has created new potential dangers for its discoverers and their descendants'. A decade later, Project Chariot promised a new harbour for Alaska. This 'sunny side of the atom' presupposed building on nature's foundations with new technologies. Emergent notions that atomic energy could improve on nature's design filtered into popular discourse surrounding even the Doom Town tests. According to one documentary, 'sagebrush and Joshua tree comes [sic] alive with activity', during the tests, as if a dead and inert desert could be brought back to life by the atom. Borrowing from Isaiah 2: 4, 'They will beat their swords into plowshares and their spears into pruning hooks,' Project Plowshare (conceived in 1957) sought peaceful applications for the bomb. Fecund atomic energy seemed eminently capable of making deserts 'bloom'. The atom posed as an invisible force fertilising vespertine plants, a desert wonder to behold as a new nuclear dawn arrived. This sense of the 'nuclear' as something spectacular was related by many eyewitnesses. One reporter described the atomic clouds over Mount Baldy 'turn to a beautiful hue, almost like a rainbow', while *Nevada Highways and Parks Magazine* labelled the Annie Test 'a gorgeous fireworks display on a gigantic scale'.[31]

Struggling to comprehend the event, news reporters frequently turned to naturalistic metaphors to explain what had happened before them. The *Las Vegas Review-Journal* reported that, 'In a civil defense trench just two miles from the blast, it was a blinding silence, the shudders of an earthquake, and then the crashing roar of furious sound and a hurricane of dust, sand and stone'. Journalist Colin McKinlay related how 'It hit us like a bolt of lightning and flying curbstone all rolled into one'. While hardly playing down the intensity of the explosions, the choice of prose constructed a natural disaster from a nuclear one.[32]

Photographic documentation of damage to Doom Town supported this line. Collapsed buildings and trailing power lines could have been easily produced by a hurricane or all manner of catastrophes. The nuclear attack looked no different from a natural disaster. A 1953 FCDA report highlighted the synergies between nuclear and natural calamities, noting that agency work could help with 'A series of natural disasters (tornadoes, floods, droughts), highlighting the peacetime need and value of civil defense'. In terms of evacuation planning and response, there was some truth to this. One 1953 pamphlet on 'emergency feeding' explained, 'In any general natural or man-made disaster – whether it be fire, flood, hurricane, tornado, earthquake or atomic, thermo-nuclear, or saturation bombing – certain emergency conditions will generally prevail'.[33]

Scholar Joseph Masco suggested that the FCDA consciously made nuclear disaster appear ordinary. Commenting on Operation Cue, Masco related: 'the film ultimately promised viewers that nuclear war could be incorporated into typical emergencies and treated alongside natural disasters such as hurricanes, earthquakes and floods'. In that way, catastrophe culture seemed relative and controllable. Doom Town propaganda complemented other attempts to 'naturalise' the atom, and thus desensitise the issue of radiation. By curtailing discussion of fallout, officials ultimately limited environmental understanding in both professional and popular circles.[34]

However, some commentators pondered the dark side of nuclear testing and projected an atom highly unnatural and dangerous. On viewing the original Trinity Test in July 1945, Programme Director Robert Oppenheimer quoted a passage from the Bhagavad Gita, 'I have become death, destroyer of worlds.' Privately, some experts voiced concerns over the threat to the environment posed by the technology of the bomb. James G. Terrill, chief of the Radiological Health Program, warned in the mid-1950s how 'Fallout from many nuclear tests is now always present in the air we breathe and the water supplies for ourselves, our animals, and our plants.'

Environmental collapse seemed plausible. A polar position started to take hold, of 'nature' versus 'nuclear' in popular dialogue. One magazine described the Annie Test as a 'Desert Holocaust.' By the 1970s, environmental protesters at nuclear installations lambasted nuclear energy as an unnatural and unwanted technology, conceptualising atomic power plants as prototype doomsday landscapes.[35]

Doom Town, meanwhile, served as a doomsday landscape, a forebear of what was to come. In one article, the *Albuquerque Journal* portrayed the town as a real location in Nevada with seventy residents killed before 'Civil defense crews moved into the broken, silent town to look with awe at its crushed homes' and carry out the dead wrapped in blankets. Doom Town was a dead place, lifeless in nature and man, an environmental omen.[36]

Atomic entertainment in the new West

Less than 100 miles from Las Vegas, Doom Town drew interesting comparison with the original sin city. An artificial structure constructed overnight and surrounded by swathes of desert, the AEC townscape resembled a micro-Vegas, a model of 1950s decadence captured in miniature. Both Las Vegas and Doom Town were viewed as places of high risk and gambling, vignettes of excess. Gambling adjectives littered popular discourse, making an impeccably planned military experiment seem like a game of chance. Referring to the unpredictable weather in the desert, the *Albuquerque Journal* related how 'it takes a gambler's jackpot luck to get off an atomic shot the size of that aimed at Survival City'. Vegas parlance rubbed off on the Nevada Test Site. An explosion was akin to hitting the jackpot. Scientific and military planning gave way to luck. The Cold War was one big poker game, with Doom Town representing just one card.[37]

Both Doom Town and Las Vegas invited the public in. As entertainment landscapes, shocks were very much on show. Like the Las Vegas Strip, with casinos constantly imploding and arising, Doom Town was blown apart then resurrected. Both sites were publically accepted as places of destructiveness, somehow capturing together apocalyptic visions, and serving as endpoints for collective American culture. In a broader, more moralistic America, these were exceptional spaces, dark and sullied. Equally, the adage 'What Happens in Vegas, Stays in Vegas,' applied to the test site:

another world of secrecy, but mushroom clouds failed to obey such artificial strictures.

Throughout the 1950s, Vegas boosters embraced the atom as a business and entertainment opportunity. Fallout was gold dust, the bomb a modish marketing tool. Consistently looking for novelty and new means to sell the city, casino owners transformed the atom into an advertising banner, attempting to turn atomic catastrophe culture into financial profit. The *Las Vegas Sun* advertised the open shot of 1955 as one of 'four major coming attractions' for the new season, the bomb show competing with a fight, a golf tournament and a Motorola convention. The atomic test was nicknamed the 'big show'. The bomb emerged as a product of 1950s kitsch, associated with Miss Atomic Bomb beauty pageants, atomic cocktails and hairdos, and parties to celebrate explosions. In the process, it became commodified and mainstream, situated as part of a wider consumer and entertainment landscape celebrating doom.[38]

The atom also made money for Vegas: tourists came to witness the atomic tests from the comfort of motels, or took buses to viewing points. 'For the first time in its history there's not a room to be had for love nor money in this gambling capital', related the *Albuquerque Journal* on the eve of Apple 2. Some 257 motels and thirty hotels were full to capacity thanks to the intersection of nuclear testing, golf and business interests. 'The dusty streets of this desert resort are clogged with traffic', the newspaper continued. As one article simply put it, 'The Boom-boom has added much to the boom in this Times Square in the Desert.' A tourist impulse connected with the atom.[39]

Some people actually went to Vegas not to gamble. In a letter to local civil defence director, Cy Crandall, test manager Carroll Tyler imparted his approval of the free trips offered to see the bomb, of the educational worth 'through them removing ungrounded fears for public safety'. Crandall, meanwhile, appreciated 'the coffee and doughnuts which were served at Mount Charleston look-out point' that also 'made a big hit' – presumably not as a big as the bomb. A fellow civil defence organiser told Crandall 'you have the most enviable assignment in civil defense work anywhere to be dealing with it in your community'. The press used the Flamingo pool for meetings while the Ancient and Honorable Society of Atom Bomb Watchers met regularly there. The bomb helped forge Vegas as more than just a gambling hotspot, it fed the city's trajectory into wider entertainment.[40]

Only occasionally was the relationship between the atom and sin city called into question. The *Las Vegas Review-Journal* explored the scenario 'If an atomic bomb was dropped on Las Vegas' in a 1953 piece. The article featured a striking composite image of a fireball above Union Depot and the Golden Nugget, and dwelt upon the vulnerable location of the city nestled between the Soviet targets of Nellis Air Force Base and the Hoover Dam. One civil defence worker wrote to Cy Crandall sharing his fears of moving to Vegas in case it was attacked, asking such questions as 'is it possible for one to dig a shelter in their back yard?' and 'is one's house protected if the air conditioning is shut off?' Vegas magnate Howard Hughes feared the bomb and dreaded the frequent shudders from testing. Practical efforts at civil defence sat uncomfortably with the 'carefree' zeitgeist of sin city. In 1955, Crandall introduced 'Operation Identification' (also known as Operation Tot-Tag), a tagging scheme for Vegas children in case of attack. 'Knowing your loved one is positively identified in the event of accident, disaster or emergency is priceless', read the publicity. The tags cost 35 cents each. Photo shoots of cute kiddies clutching dog tags were seen to 'make for great pictures', but ultimately how comfortable was the city with the bomb? Practically, atomic tests often led to disappointment. Expectations were dashed at ground zero by weather interruptions and misfires. At one civil defence camp, a disappointed individual lamented: 'the only slot machine here is the turnstile into which goes a silver dollar each time someone enters the cafeteria'. Vegas encouraged expectations beyond even the power of atomic technology to deliver. Situated as an extension of the broader Vegas landscape, Nevada Test Site, and with it Doom Town, were under constant pressure to perform.[41]

As an entertainment landscape, Doom Town also interfaced with Hollywood. Ground zero served as a convenient starting point for a range of horror and disaster stories in the 1950s. Ideas about mad scientists, radioactivity and mutation ran amok in science-fiction prose. With little to go on other than Hiroshima and Nagasaki, imagination fed the public conception of nuclear landscapes. The movie *Them!* (1954), released between the two open tests, showed giant ants exiting the first atomic explosion in New Mexico's desert, the big insects causing havoc to the local populace. Realism seemed a distant prospect. To a degree, Doom Town was deliberately designed by the FCDA to dispel the growing tide of atomic myths, and show for the first time the tangible effects of nuclear attack on main-street America. The media hoped for something truly entertaining, however. Television producers cried out for an exciting documentary and shots to rival any channel. Journalists complained about the bombs 'muffing their cue', casting Annie and Apple 2 as poorly performing actors in a nuclear theatre. Bearing a remarkable similarity to a Hollywood film set, Doom Town itself somewhat fitted- the Hollywood image. Thanks to the ghostly mannequins, the *LA Times* dubbed it 'terror town'. The rise and fall of Doom Town did little to uproot new atomic culture. In fact, archival footage of Test Annie was quickly co-opted into the Hollywood movie-making machine. One film recreated the experiment in new dimensions, advertising for *Doom Town* (1953) promising viewers the chance to 'See the Atomic Bomb in 3D', to witness for the first time the 'Death of a City by Atomic Destruction'. The 'light comedy' *Atomic Kid* (1954), featuring Mickey Rooney, told of a uranium prospector getting caught in Doom Town as the bomb goes off, narrating his exploits after the bomb (complete with glowing eyes). *Atomic Kid* utilised real footage of Annie but the bizarre nuclear comedy bombed.[42]

The nu-king of suburbia: 'the city that was built to live for but one day'

The nuking of Doom Town provided a powerful doomsday image. With the construction of suburbia in the desert wastelands of Nevada, the bombs were hitting something valuable for the first time. Rather than eliminating sprawling sagebrush, tests Annie and Apple 2 threatened a greater definition of life beyond the desert. The FCDA effectively tapped contemporary fascinations with suburbia, post-war family values and mass consumption, with the warning that all these things were now at risk with the bomb. In a practical sense, with the construction of Doom Town 1, officials hoped to gauge the effects of a nuclear blast on small-town or suburban America. In response to press releases billing Annie as 'the first time American homes would be tested', the *Las Vegas Review-Journal* labelled Doom Town a 'laboratory city' where lessons could be learnt to 'prevent real atomic tragedy'. The newspaper hoped that Doom Town 'will supply the formula for averting wholesale disaster', as if scientific magic gleaned from the detritus of a wrecked townscape could somehow prevent Armageddon.[43]

Scientific information nonetheless paled alongside the propaganda purposes of Doom Town. The FCDA hoped that the two open tests would bring home the dangers of nuclear attack and rally civilians around the concept of civil defence. Reportage mustered a sense of anticipation over the tests. One newspaper related, 'the nation waits for the fate of a tiny village, wondering what will happen. Wondering if it can happen to his town, his home, his loved ones.' The dress rehearsal for Armageddon involved mannequins and civil defence workers on site, everyone else able to watch on television. Part of civil defence preparedness, immersion in the mock disaster proved a valuable educational experience. Doom Town was about bringing doomsday home, making it real, pushing people into action. Allied to Duck-and-Cover school drills and the construction of home shelters, the Doom Town experiment facilitated the 'domestication' of the bomb. The atom was introduced to suburbia, to ordinary life, situated alongside the average American home. Camera operators broadcast doom across the nation's television sets. One programme noted its aim to 'make the threat of the mushroom cloud a dark reality across the nation'.[44]

The official *Doorstep* television documentary made the most of this narrative of impending doom. Opening with a scene 'typical of the American family at home', with 'children at play unaware of approaching disaster', the programme projected a sense of temporary calm and normality. Then the 'drama of survival' began with the bomb, throwing life into chaos. Everything fell apart and destruction reigned. This fall from grace for the American town, a classic environmental tale of declension, was later picked up by Rachel Carson in *Silent Spring* (1962).[45]

Making the disaster eminently real entailed consummate licence. The FCDA aimed for a middle-class street aesthetic that reflected contemporary architectural design. Doom Town perfectly intersected with the rise of American suburbia. The fast, standardised creation of both Doom Towns replicated the Fordist assembly line model practised at Levittown, New York, just a couple of years earlier. The look of the Doomsday Drive houses, with their fresh white paint, simple lines, and wooden contours, closely resembled the fashionable prairie house of the period. The new suburban West, flourishing at places such as Lakewood, Los Angeles, played out in microcosm at Nevada Test Site. Doom Town signified an atomic-styled Levittown of the 1950s.

'Complete with the most up-to-date furnishings from refrigerators to kitchen stoves, even lace curtains', the houses on Doomsday Drive were filled to the brim with consumer products of the 1950s. Doom Town was like entering an atomic-themed department store or bone-shaped shopping mall of the period, featuring a collection of new and unused goods, and amounting to a storehouse of consumer capital. Viewed alongside the kitchen debate of Nixon and Khrushchev, the Nevada townscape represented a successful and victorious American landscape of material abundance. 'Rockets and refrigerators were equally weapons on the Cold War battlefield', home appliances enlisted as weapons of consumer power.[46]

Mannequins brought the Doom Town story alive in media and press. They provided a sense of shape and colour to an invisible threat, their twisted bodies demonstrative that nuclear blasts could hurt people as well as buildings. The *Las Vegas Review-Journal* detailed snapped telephone poles and crumpled sheds, 'A mannequin mother blown to bits as she spooned baby food to her department store dummy infant', and 'A baby mannequin the size of a your three-year old lying in a crib under needle-fine glass that once had been a window'. Everyone and everything seemed at threat, the atom symbolic of total environmental destruction. Newspapers exploited this alarming narrative of catastrophe culture. One article carried the worrying headline 'expendable! The entire population of Doom Town II, USA.'[47]

The power of the mannequins also rested with their status and what they stood for. First and foremost, the Doom Town dummies represented real Americans facing the spectre of nuclear warfare. On record, Rear Admiral Robert W. Berry related how, 'It is indeed fortunate that we can examine these mannequins, which serve as proxies for our citizens'. The mannequins served the nation by collecting data for the Cold War effort. Alongside photographs of the Doom Town families huddled under stairwells, papers reported how 'Mannequins tell graphic defense story', granting the props an unusual level of agency. The Nevada mannequins were real people, given life, given purpose, by test observers. Masco related how 'viewers were invited to think of themselves as mannequins caught in an unannounced nuclear attack'. 'Mannequins thrown about, clothing cut, plaster bodies potmarked by flying glass' were Americans unprepared for nuclear attack. The narrator for the *Operation Doorstep* documentary highlighted how one mannequin household at Doom Town was ultimately 'Unprepared', 'They did not take shelter'. The experiments represented a call to action for real Americans to make home shelters, 'or will you, like a mannequin, just sit and wait'.[48]

Sporting department-store attire, 1950s sedans outside their suburban houses, and the latest kitchen appliances inside, the Doom Town mannequins represented the 1950s nuclear family at home. Released photographs documented one mother with her child and another whole family watching television in the living room. At Doom Town II, one group residing on Doomsday Drive was nicknamed the Darling family. The *Las Vegas Review-Journal* ran a novelty piece around a fictitious interview with the 'sole survivor' of the explosion, 'Junior Darling', 'orphan when the remainder of his family was wiped out in the holocaust'. The mannequin child explained that the death of his family was due to poor choice of shelter, adding 'I'll bet Superman or Davy Crockett would have been scared'.[49]

But the cult of the mannequin exceeded their status as nuclear family avatars. The more macabre images of destruction and decay captured the most public attention. Marvin Miles for the *Los Angeles Times* waxed lyrical over 'a headless soldier, a fully uniformed waxed model, lying sprawled in the desert sage, a grotesquely torn trooper ripped by every Yucca Flat explosion since the beginning of the 1953 test series'. Billed as 'veterans of the atomic explosion', the 'Maimed mannequins of Yucca Flat' (also known as 'Doom Town dummies') became celebrities of Armageddon. Curiosities or freaks of the atomic age, the dummies toured Los Angeles, occupying a three-day exhibit at Pershing Square in April 1953 just a fortnight after the test. They posed alongside an F-84 fighter. The *Las Vegas Review-Journal* lamented how 'the plane draws a larger crowd than the "people who came out of an atomic blast"'. The tour was a pageant to Armageddon. The radioactive homeless entertained as street theatre and spectacle.[50]

The presence of motor cars at both Doom Towns completed the picture of suburbia. Parked outside residences, the range of new saloons and estate cars cemented a sense of a post-war consumer economy and technology. The vehicles symbolised a route out of Armageddon on several fronts by offering fast transport away from ground zero, shelter from the blast, and adequate storage for supplies. As the FCDA booklet, *4 Wheels to Survival* (1955), noted: 'Your car helps you move away from Danger', 'Your car can be your shopping center' and, thanks to Doom Town, 'tests under an actual atomic explosion in Nevada proved that modern cars . . . give a degree of protection against blast, heat and radiation'. The booklet even featured a section on conserving petrol in case supplies run out, a precursor to environmental worries over oil production. In conclusion, the FCDA announced that 'the typical American automobile would hold its own', and that the 'car is a pretty good shelter'. This propaganda element reflected the line of official reports, with one 1953 FCDA missive hoping that Annie results would filter into the 'future design of cars' and accumulated evidence that 'Almost without exception, a car that would start and run before the March 17th blast could be started and driven after the blast'. Good news for D. L. Johnson of Johnson Bros Chevrolet, Texas who kindly furnished a 1949 Chevrolet Sedan for experimentation, with one request, 'I would like to have my car regardless of the condition it might be in, even is [sic] I have to send a truck to haul it back to Dallas.'[51]

The overriding impression for visitors to Doom Town was the scale of authenticity and detail lavished on the experiment. Doom Town 2 featured paved streets, white goods and tablecloths. The FCDA constructed a realistic townscape with an unbridled sense of immersion for civil defence workers and military professionals. It was equally fake, fantastical and like a Hollywood film set. Such attributes made Doom Town an interesting counterpart to another fantasy world located in the nearby state of California. Disneyland opened in Anaheim just two months after Doom Town 2 was blown to pieces. Walt Disney's fantasy theme park boasted its own facsimile main street populated by movable mannequins (or in Disney parlance, audio-animatronic figures). It equally indulged in 1950s abundance and endorsed the iconographies of consumption. The same visual symbols (or cultural signifiers) of the period could be found at both places: dream kitchens, high technology, and machine automation. Both landscapes embodied geographic simulations of the new middle-class American dream of cleanliness, whiteness (in people and products), social order and capital. There was one crucial difference, however. At Disneyland, Walt Disney aimed to provide a totally immersive architecture of reassurance. The fantasy park served as a buffer zone from anxieties of the period, a comfortable world seemingly untouchable by the perils of the Cold War, a perfect escape. Doom Town, too, boasted an architecture of reassurance in its middle-class order and fresh facade. But its destruction shockingly revealed that any hope of escape from the bomb was a fallacy, a cartoon fantasy.

Doom Town highlighted what was at stake in the Cold War: cars, clothes, luxuries, the nuclear family, and the American way of life. And blowing up Doom Town was like blowing up Disney's Main Street, ruining the fantasy.[52]

In raising the spectre of nuking suburbia, Doom Town complemented other civil defence propaganda of the period. The FCDA produced the film *Disaster on Main Street* in 1953. The documentary *The House in the Middle* (1954) broadcast the need to keep the American home spick and span to limit the effects of firestorms in nuclear attack. Behind the gloss, and nuclear spin, resided a strong message to uphold 1950s values: a clean, orderly house and home life. Reflecting budgetary constraints and the need for corporate ties, the National Paint, Varnish and Lacquer Association produced the film. The FCDA did, nonetheless, take such issues seriously. The Upshot–Knothole series included a *Study of Fire Retardant Paints* along with fire-impact studies on the Doom Town houses.[53]

The Doom Town experiment also preached the value of conformity, both in terms of venerating traditional family values but also by enforcing gender roles, too. Frequently it was the prettified female mannequins in a domestic setting that received the most coverage. One photograph showed Lydia Hurst, assistant to the FCDA director, setting the table for Mrs Mannequin and daughter, with the caption of the family being 'in for a rough meal'. The blissful calm of the 1950s domestic goddess was about to be rudely interrupted by the atomic age.[54]

The official documentary produced for Operation Cue followed a female reporter, Joan Culler, on her journey to ground zero. Culler promised to see the test, 'through the eyes of the average American man and woman', inculcating a level of trust in the viewer by situating herself as 'one of us'. Culler was particularly impressed by the level of authenticity of Doom Town, homes 'complete in every detail', with 'the things we use in everyday lives', from home fabrics to food tins, about to be tested. Culler fondly referred to the mannequins as 'Mr and Mrs America'. The performance proved highly orchestrated, Culler a vessel for standard FCDA and military messages about disaster planning. Tellingly, an official-sounding male narrator provided the answers to all her questions.[55]

Another minor celebrity of the Doom Town experiment, Mrs Jean Fuller, a Los Angeleno and FCDA volunteer, attracted press attention. The emphasis was again on the unique female perspective, with Fuller consciously offering her 'own story' of Operation Cue. Standing with seven other females in trenches close to ground zero, notably 'the first time women have been that close to an atomic explosion', Fuller detailed how 'the normal feminine excitement prevailed amongst us all, but I didn't feel that my life was in danger'. Fuller found the test 'terrific, interesting and exciting', referring to Dante's Inferno as her literary device. She also compared the aftershock with a natural disaster, 'being a Californian, it reminded me of a good stiff earthquake'. Her verdict: 'that women can stand the shock and strain of an atomic explosion just as well as men'. Others found the tests more troubling. Violet Keppers from Missouri related in an article for *Parade* magazine (entitled 'A Housewife watches the A Bomb . . . I saw a stairway to hell') that the event simply 'made me cry'. Rae Ashton from Utah felt 'scared', and noted how 'it was hardly a household teaspoonful of a bomb'.[56]

The nuclear tests reinforced rather than challenged the status quo. The mannequin man in the grey flannel suit appeared perfectly suited to being blasted at ground zero. Doom Town touched on the intriguing relationship between conformity and the bomb, of the desire for stability in times of danger. As Susan Sontag related, 'For we live under continual threat of two equally fearful, but seemingly opposed, destinies: unremitting banality and inconceivable terror.' Doom Town captured both futures.[57]

Against such terror, the FCDA promoted American faith in survivability of nuclear attack based around historic notions of individualism. Predating triumphalist nuclear films such as *Red Dawn* (1984) by some three decades, the agency pushed a victory narrative based on self-help. There was a clear survival message here. After Apple 2, the FCDA released information that deemed shelters and cars worthwhile ways of avoiding death. The *Operation Cue* documentary suggested that with 'many lessons learnt', a 'plan for the survival of homes, our families and our nation in the nuclear age', arose from the debris of Doom Town. Occasionally, a sense of a survival spirit did win out, the line 'out of a makeshift bathroom shelter came two cheerful, tail wagging witnesses for survival', testament to canine resilience at the very least. Others commented on the resilience of the buildings. But could Americans win out in the Cold War?[58]

Lingering visions

The FCDA succeeded in raising public consciousness of nuclear attack through the Doom Town experiments. The open tests were tied to a fresh push on civil defence leafleting and publicity, part of a new Cold War patriotic 'call for heroes' or, as the *Las Vegas Sun* put it, 'America needs her heroes now'. The Doom Town experiment engaged the popular imagination. As Jean Baudrillard recognised, simulation remains a powerful force in America. The linkage of the iconic mushroom cloud with a fake suburbia was truly arresting. As the *Operation Cue* narrator claimed, Doom Town tied 'home, family, nation' to testing. Doom Town captured the zeitgeist of 1950s America: a landscape of consumption, conformity and domesticity frozen in time when the bomb went off. The iconic image of the mushroom cloud, symbol of the Cold War, now had something below it. Watched my millions, the *Doorstep* documentary amounted to a twisted nuclear family sitcom, ending with the classic but haunting line, 'Will you, like a mannequin, just sit and wait'. Doom Town represented a worthwhile fight on apathy by recourse to a combination of shock tactics and imagination. The FCDA recognised the importance of fitting a new environmental danger within an 'existing and familiar' landscape of suburbia. Previously, the atom had seemed remote and distance, something 'outside' and other. The new proximity of peril proved potent.[59]

The FCDA forged a powerful doomsday image at Nevada Test Site but the Doom Town experiment exhibited significant flaws. Part of a broader strategy to frame the public mindset and dictate catastrophe culture, the open test series failed to shift attitudes towards civil defence on a fundamental level. A maximum of four million volunteers enrolled for civil defence programmes in the 1950s, barely enough for a nuclear home front. The FCDA failed to 'balance' terror with messages of survival. Doom Town joined a range of compelling images of horror, from decimated Japanese cities to Soviet totalitarianism but people needed motivation other than fear. As McEnaney related, 'unpleasant images of bombs dropping from the sky was precisely what the FCDA had to conjure up in order to activate citizens, but such scenarios had to be matched with equally gripping scenes of survival and triumph'. The big question of nuclear survival remained perilously unanswered at Nevada Test Site. Apple and Annie tests failed to convince anyone about the best course of action in nuclear attack. Even the most sheltered of mannequins and outbuildings were damaged by fallout. Most press attention focused on the destructive power of the atom rather than on any survival motifs. Headlines such as 'Most residents of Blasted Town Would Have Died', hardly inculcated optimism in the merits of civil defence or the concept of survivalism. Ultimately, the 'blowing up of suburbia' represented too perfect a nightmare scenario. Images of distorted dummies and collapsed electricity pylons highlighted the futility of nuclear war rather than a winnable Cold War. Operation Doorstep brought death to the front porch of 1950s America. The open tests were polite invitations to watch doomsday play out.[60]

This atomic doomsday scenario would go on to become the dominant image of environmental collapse in the twentieth century. It would become linked to scientific and technological ideas about nuclear winter, genetic decay, birth mutations, cancer, endangered species loss, modern pollution and fractured chains of life. It would reoccur in various guises in other environmental crusades, from *Silent Spring* (1962) to Love Canal. In popular fiction and film, narratives of death and destruction would frequently call on the atomic motif and nuclear plot device. Movies from *Planet of the Apes* (1968) to the *Hills Have Eyes* (1977), and *Mad Max* (1979) to *Testament* (1983) went on to explore post-apocalyptic wastelands, creating new kinds of Doom Towns and test sites. For most Americans, the notion of environmental doomsday came to feature a radioactive symbol.[61]

Notes

1. *Bulletin of the Atomic Scientists* (September 1953).
2. Dwight Eisenhower, Atoms for Peace speech (8 December 1953).
3. Donald Worster, *Nature's Economy: A History of Ecological Ideas* (Cambridge: Cambridge University Press, 1994 [1977]), pp. 342–3.
4. For history of NTS, see Terence R. Fehner and F. G. Gosling, *Origins of the Nevada Test Site* (Las Vegas: Department of Energy, 2002), Richard L. Miller, *Under the Cloud: The Decades of Nuclear Testing* (Woodlands, TX: Two-Sixty Press, 1991), A. Costandina Titus, *Bombs in the Backyard: Atomic Testing and American Politics* (Reno: University of Nevada Press, 1986).
5. FCDA Advisory Bulletin 'Disaster Series' 1953 (Nevada State Museum, Las Vegas, NSM MS28 1/3); For information on the FCDA, see Dee Garrison, *Bracing for Armageddon: Why Civil Defense Never Worked* (Oxford: Oxford University Press, 2006), Andrew D. Grossman, *Neither Dead Nor Red: Civilian Defense and American Political Development during the Early Cold War* (New York: Routledge, 2001), Laura McEnaney, *Civil Defense Begins at Home: Militarization Meets Everyday Life*

in the Fifties (Princeton: Princeton University Press, 2000). For civil defence films in particular, see Joanne Brown, '"A is for Atom, B is for Bomb": Civil Defense in American Public Education, 1948–1963', *The Journal of American History* 75/1 (June 1988), pp. 68–90, Joseph Masco, 'Target Audience', *Bulletin of the Atomic Scientists* 64/3 (July/August 2008), pp. 22–31.

6. John C. Clark, Report of the Deputy Test Director, 'Operation Upshot–Knothole' 1 March 1953 (Nuclear Testing Archive, Las Vegas, NV0306227) pp. 19, 22, 23. For general details of Annie/Upshot–Knothole, see: Memo from R. L. Southwick, Assistant Chief, Public Information Service, 'Draft Information Plan for "Open" Shot Operation Upshot–Knothole,' 12 February 1953 (NV0404682); J. Massie et al., 'Shots Annie to Ray', 14 January 1982 (NV0018987); DNA, 'Fact Sheet: Operation Upshot–Knothole' 11 January 1982 (NV0760121).

7. Fletcher Knebel, 'We're wide open for Disaster', *Look* 15/5, 27 February 1951 (Nevada State Museum MS28 Clark County Civil Defense Records 1/27); also see FCDA 'Annual Report for 1953' (NSM MS28 1/20).

8. Bob Considine, 'Great Atomic Bomb Sale Due Today', *Los Angeles Examiner*, 17 March 1953; Bob Considine, 'Everything's Set for Big Atomic Test', *Las Vegas Review-Journal*, 17 March 1953; Civil Defense Films, *Operation Doorstep* documentary 1953 (NV0800033); Atomic Testing Museum exhibit, Las Vegas.

9. For an example of the Survival City moniker, see, 'Doom City Thoroughfare Damaged Severely', *Las Vegas Review-Journal*, 6 May 1955; A decade later, the same town name surfaced on the other side of the Atlantic when the British Royal Air Force converted a disused military station off the Cumberland Coast into a nuclear-related civil defence training centre. A promotional film for the RAF noted how the 'dead town' was doing a 'live job'. Pathe News, *Cumberland Doom Town*, documentary (undated).

10. Massie et al., 'Shots Annie to Ray'; Clark, Report of Deputy Test Director, p. 49; Fallout: Lt Col. N. M. Lulejian 'Radioactive Fall-out from Atomic Bombs', 1 November 1953 (NV0755905); Tumbler documentary (1952).

11. 'Troops in Foxholes View "Greatest Show on Earth"', *Las Vegas Review-Journal*, 17 March 1953; Robert Bennyhoff, 'Newsmen Set for Ringside Blast Seat', *Las Vegas Sun*, 17 March 1953; 'Televiewers to See A-Blast Havoc', *Las Vegas Review-Journal*, 11 March 1953.

12. Defense Nuclear Agency, 'Shot Apple 2: A Test of the Teapot Series' 25 November 1981 (NV0017806) pp. 9, 52; Bob Considine, 'Variety of Objects to be Subjected to A-Bomb in Civil Defense Test', *Albuquerque Journal*, 24 April 1955. Also see: Nevada Test Organization, 'Background Information on Nevada Nuclear Tests' 15 July 1957 (NV0403243).

13. 'Mock City Ready for Atomic Test', *Los Angeles Times*, 24 April 1955; CBS 'adventure': 'Camera Tour of Atomic Proving Grounds Set Tomorrow on KLAS', *Las Vegas Sun*, 23 April 1955; Hillman Lee: 'Penney's to "Populate" Yucca Flat's Doom Town', *Las Vegas Review-Journal*, 24 April 1955; Considine, 'Great Atomic Bomb . . .'; 'American Community': DNA 'Shot Apple 2', p. 52. Also see Archie Teague, 'Doom Town to Show Way to Survival', *Las Vegas Review-Journal*, 22 April 1955 and Considine, 'Variety of Objects . . .' Considine noted how, 'The "guinea pigs", human and otherwise, range from high ranking generals to rookie air wardens, from Patton tanks to cans of peas, from rugged Marines to sensitive nuclear physicists, from fully furnished houses with gas, water and electricity in working order to a bundle of diapers.'

14. Considine, 'Variety . . .'.

15. 'Newsmen join GIs' and 'ready': Colin McKinlay, 'Tanks Shield Observers from Atom Blast Effects', *Las Vegas Sun*, 6 May 1955; 'A nuclear weapon': Teague, 'Doom Town . . .'; 'It's "Out of This World" at Atom City', *Oakland Tribune*, 24 April 1955; 'premier': John F. Cahlan, 'Expensive "Premier" Set Sunday', *Las Vegas Review-Journal*, 24 April 1955; Marvin Miles, 'Test Village Awaits Effect of Atom Blast', *Los Angeles Times*, 25 April 1955.

16. Anthony Leviero, 'Atom Blast Wait Angers Officers', *New York Times*, 30 April 1955;

Alan Jarlson, 'Inside Las Vegas', *Las Vegas Sun,* 29 April 1955; Frank La Tourette, 'An A-Bomb Can Muff a Payoff TV Cue', *The Quill,* July 1953; Jake Freedman, The Sands, 'To Our Sands Reservations' April 26, 1955 (University of Nevada, Las Vegas, UNLV SC Sands Collection 49/3); Freeman Company Public Relations report for Sands, 11 July 1955 (UNLV SC 49/3).

17. AEC, 'Eighteenth Semiannual Report', July 1955, p. 2 (NV0727884); DNA, 'Shot Apple 2', p. 52; Fallout: Interview with John Auxier, Oak Ridge, 13 April 1979 (NV0702882) and DNA, 'Operation Teapot' 1 January 1955 (NV0767828); Leininger quoted in 'Doom City Main Thoroughfare Damaged Severely', *Las Vegas Review-Journal,* 6 May 1955.

18. McEnaney, pp. 11, 30.

19. Val Peterson, 'Panic: The Ultimate Weapon' (1953) (NSM, MS28 3/1).

20. September 1951 report cited in Advisory Committee on Human Radiation Experiments, Staff Memoranda: Atomic Veterans, 8 September 1994 (NV0759709), p. 4; 'fear response': P. Broudy, 'Exposure Suffered by Atomic Veterans', 18 July 1994 (NV0750329), p. 3.

21. US Government, *Survival Under Atomic Attack* (1950), p. 2 (NSM, MS28 1/27)

22. Guy Oakes, *The Imaginary War: Civil Defense and American Cold War Culture* (New York: Oxford University Press, 1994), p. 34.

23. FCDA, *Atomic Attack: How to Protect Yourself* (undated; early 1950s) (NSM MS28 3/37).

24. Hiroshima, for example, Miles, 'Test Village . . .' and 'Specter of Ruin Stalks A-Test City', *Oakland Tribune,* 25 April 1955; Model T: 'Specter'; Pearl Harbor: see Oakes, pp. 21, 44.

25. Scott Kirsch, 'Watching the Bombs Go Off: Photography, Nuclear Landscapes, and Spectator Democracy', *Antipode* 29:3 (1997), p. 229; Tom Vanderbilt, *Survival City: Adventures Among the Ruins of Atomic America* (NY: Princeton Architectural Press, 2002), p. 39; Fehner, pp. 39–50; Editorial: 'An Awesome Thing', *Wyoming Eagle,* 19 March 1953; AEC, *Report of Committee on the Operational Future of Nevada Proving Grounds,* 11 May 1953 (NV0720368), pp. 4, 3.

26. See Patricia Nelson Limerick, *The Legacy of Conquest: The Unbroken Past of the American West* (New York: Norton, 1987), Kevin Fernlund, ed., *The Cold War American West* (Albuquerque: New Mexico University Press, 1998), and Bruce Hevly and John Findlay, eds, *The Atomic West* (Seattle: University of Washington Press, 1998).

27. Vanderbilt p. 25; Wallace Stegner, *Mormon Country* (1942), p. 33; Valerie Kuletz, *The Tainted Desert: Environmental and Social Ruin in the American West* (New York: Routledge, 1998). Also see: Rebecca Solnit, *Savage Dreams: A Journey into the Landscape Wars of the American West* (New York: Vintage, 1994).

28. Michael Amundson, 'Home on the Range No More: The Boom and Bust of a Wyoming Uranium Mining Town, 1957–1988', *Western Historical Quarterly* 26/4 (winter 1995), p. 490 and Michael Amundson, *Yellowcake Towns: Uranium Mining Communities in the American West* (Boulder: University Press of Colorado, 2002); 'one day': Cahlan, 'Expensive . . .'; 'A-Boom Town is a Doomtown', *Los Angeles Examiner,* 26 April 1955; 'ghost town' and 'mule train': Jarlson, 'Inside . . .'; Daniel Lang, 'Blackjack and Flashes', *New Yorker,* 20 September 1952; 'Newsmen, Observers Wade Through Doom Town Today', *Las Vegas Sun,* 6 May 1955; Archie Teague, 'Dust Piles Over "Dead" In Nevada's Doom Town', *Las Vegas Review-Journal,* 8 May 1955.

29. William Riebsame, ed., *Atlas of the New West: Portrait of a Changing Region* (New York: Norton, 1997), 'Ugly West' section, p. 134.

30. Miles, 'Test Village . . .'; *Washington Post* quoted in McEnaney, p. 55.

31. AEC, *Sixth Semiannual Report of the Atomic Energy Commission,* July 1949, p. 114 (NV0727863); 'Utahns See Horizon Painted by Blast', *Los Angeles Examiner,* 17 March 1953; *Operation Doorstep* documentary.

32. 'Doom City'; McKinlay, 'Tanks Shield . . .'.

33. FCDA, *Annual Report*, 1953, p. 68 (NSM MS28 1/20); Department of Defense and FCDA, *Emergency Mass Feeding Instructor Course* (Washington, DC: Government Printing Office, 1953) (NSM MS28 2/9).

34. Masco, p. 29.

35. J.G. Terrill, "The Nature of Radioactive Fallout" Technical Presentation for the Joint Committee on Atomic Energy Hearings (24 May 1957), p. 9 (NV0407635).

36. 'Most Residents of Blasted Town Would Have Died', *Albuquerque Journal*, 7 May 1955.

37. 'Gambler's Jackpot Luck Needed for Big A-Blast', *Albuquerque Journal*, 5 May 1955. See also Ken Cooper, '"Zero Pays the House": The Las Vegas Novel and Atomic Roulette', *Contemporary Literature* 33/3 (1992), pp. 528–44.

38. 'Half Dozen Big Attractions Bring Major Headaches', *Las Vegas Sun*, 21 April 1955.

39. 'Standing Room All That's Left in Las Vegas', *Albuquerque Journal*, 25 April 1955; Earl Wilson, 'Columnist Wilson Finds This Is Boom-Boom Town', *Las Vegas Review-Journal*, 24 April 1955.

40. Carroll Tyler, AEC test manager, to I. R. Cy Crandall, director, City-County CDA, 28 May 1952 (NSM MS28 1/1); I. R. Cy Crandall to Miss Clara Hogg, Red Cross, 13 June 1952 (NSM MS28 1/1); Frank S. Carroll, FCDA, to Cy Crandall, 14 April 1953 (NSM MS36 1/15).

41. Jeff McColl, 'If an Atomic Bomb was Dropped on Las Vegas', *Las Vegas Review-Journal*, 31 May 1953; D. Duvall, Murray, Utah, to Cy Crandall undated, April 1955 (NSM MS28 2/10); Ralph Vartabedian, 'Howard Hughes and the atomic bomb in middle Nevada', *Los Angeles Times*, 28 June 2009; Freeman Public Relations, 'Operation Tot-Tag' and School Memo (including ideas such as sending youngsters to see blasts and essay-writing competitions ('How the Atom Could Build a Better America and World')), O. J. Toland, 'Operation Identification' (see NSM MS28 2/10); 'It's "Out of This World" . . .'.

42. "Test Village..."; Tourette, "An A-Bomb..."; More on nuclear movies, see Joyce Evans, *Celluloid Mushroom Clouds: Hollywood and the Atomic Bomb* (Boulder: Westview Press, 1998), Toni Perrine, *Film and the Nuclear Age: Representing Cultural Anxiety* (New York: Garland, 1998), Jerome Shapiro, *Atomic Bomb Cinema* (New York: Routledge, 2002).

43. Teague, 'Dust Piles . . .'.

44. Miles, 'Test Village . . .'; *Operation Doorstep* documentary.

45. *Operation Doorstep* documentary; See 'A Fable for Tomorrow', in Rachel Carson, *Silent Spring* (Boston: Houghton Mifflin, 1987 [1962]), pp. 1–3.

46. Miles, 'Test Village . . .; Vanderbilt, p. 46.

47. Teague, 'Dust Piles . . .'; 'Expendable! The Entire Population of "Doom Town II, USA"', not sourced, 26 April 1955 [held at DOE newspaper collection, Las Vegas].

48. Berry quoted in 'Mannequins Tell Graphic Defense Story', *Los Angeles Times*, 1 April 1953; Masco p. 28; *Operation Doorstep* documentary.

49. 'Editor Interviews Survivor', *Las Vegas Review-Journal*, 7 May 1955. Also see 'Atom Blast Razes "Town"; None Hurt', *Los Angeles Examiner*, 6 May 1955.

50. Miles, 'Test Village . . .'; 'Mannequins Play Second Fiddle to F-84', *Las Vegas Review-Journal*, 1 April 1953.

51. FCDA, *4 Wheels to Survival*, US Government Printing Office, 1955 (NSM MS28 3/52); FCDA 1953 report excerpts (NSM Patricia Lee, Civil Defense, MS36 1/4); D. L. Johnson, Dallas to Cy Crandall, Las Vegas, 31 March 1953 (NSM MS36 1/4).

52. More on Disney, see Karen Jones and John Wills, *Invention of the Park: From the Garden of Eden to Disney's Magic Kingdom* (Cambridge: Polity Press, 2005), Janet Wasko, *Understanding Disney: The Manufacture of Fantasy* (Cambridge: Polity Press, 2001).

53. *The House in the Middle* documentary (FCDA, 1954); *A Study of Fire Retardant Paints* (Miller, 1953), cited in *Operation Upshot-Knothole March–June 1953: Summary Report of Technical Director*, 31 March 1953 (NV0306225).

54. 'Doom Town II Ready for Tuesday's A-Blast', *Las Vegas Review-Journal*, 25 April 1955.
55. Civil Defense Films, *Operation Cue* (1955) documentary (NV0800033).
56. Jean Wood Fuller, 'L.A. Woman in Trench at A-Blast', *Los Angeles Times*, 6 May 1955; Violet Keppers, *A Housewife watches the A Bomb . . . I Saw a Stairway to Hell*, pamphlet for Missouri Civil Defense Agency (NSM MS36 1/ 25); 'Just a 'Teaspoon'' Bomb Still Hard to Believe', *Albuquerque Journal*, 18 March 1953. Also see 'Women Practice Crouch for A-Bomb Shock', *Los Angeles Times*, 27 April 1955.
57. Susan Sontag, 'The Imagination of Disaster', (1965) taken from *Against Interpretation* (London: Vintage, 2001), p. 224.
58. *4 Wheels to Survival*; *Operation Cue* documentary; Elton C. Fay, ''Death' and Destruction in Atom-Blasted Town', *Las Vegas Review-Journal*, 6 May 1955.
59. Dorothy Shaver, 'Editorials: Civil Defense Action Heroism With Heroes', *Las Vegas Sun*, 24 April 1955; Jean Baudrillard, *America* (London: Verso, 1988 [1986]); *Operation Cue* documentary.
60. McEnaney, p. 29; 'Most Residents . . .'
61. Carole Gallagher, *American Ground Zero: The Secret Nuclear War* (New York: Random House, 1993), Richard Misrach, *Bravo 20: The Bombing of the American West* (Baltimore: Johns Hopkins University Press, 1990). See also Mike Davis, in Valerie Matsumoto and Blake Allmendinger, eds, *Over the Edge: Remapping the American West* (Berkeley: University of California Press, 1999), pp. 339–69.

CHAPTER FOUR

Chemical Dystopia and *Silent Spring*

Paris Green helped keep American apples green. Around the turn of the twentieth century, horticulturalists regularly sprayed the pesticide on their apple trees. The coating ensured less damage from pests and brought more produce to the market. Dating back to 1867, Paris Green first proved effective as a weapon against rats inhabiting Parisian sewers. It also found favour as a colour palette for painters Paul Cézanne and Vincent Van Gogh. Cézanne's famous 'Green Apples' was dusted in the pigment. But, as a mix of lead and arsenic, the pesticide proved a potent poison, and probably contributed to the illnesses of both Cézanne and Van Gogh.

The deployment of pesticides in the United States was very much tied to the survival of the agrarian nation in the twentieth century. Pesticides promised greater productivity and a lifeline for farmers struggling to make profit in a harsher economic climate. Tied to professionalisation, new techniques of agriculture, commercialisation and the scientific revolution in farming, pesticides were among the new technological armaments of the modern agriculturalist. Sprayed across crop fields, the chemical agents kept pests such as the boll weevil at bay. They symbolised a quest for maximum yield and perfect, factory-like produce. In the 1940s and 1950s, a fresh range of pesticides hit the market. Chemical agents, such as DDT, were marketed as miracle products, capable of revolutionising the average American farm. They also promised to wipe out all kinds of home and garden pests from the gypsy moth to the mosquito. The 1950s epitomised the golden age of pesticides.

In 1962, Houghton Mifflin published *Silent Spring* by nature writer Rachel Carson. Carson had written three books on marine life but *Silent Spring* did something new – it challenged the use of pesticides in America. The book became an instant bestseller, the target of media chatter, chemical industry propaganda, and government investigation. The safety of pesticides came under review and President Kennedy set up a presidential commission to investigate the danger of DDT. The publisher placed advertisements in newspapers telling how 'The whole country is talking about Silent Spring'.[1]

The perfect doomsday scenario

Much of the impact of *Silent Spring* rested on the manufacture of a perfect doomsday scenario within its pages, the forwarding of a powerful apocalyptic image. Carson achieved this in the opening chapter, 'A Fable for Tomorrow', a moving story of environmental collapse. In early plans for the book, Carson envisaged the chapter as a summary of 'the entire thesis and [will] actually stand for the book in miniature'. In fewer than four pages, Carson came to map out a frightening environmental dystopia. She related the end of the world by chemical spray. Carson provisionally titled it 'The Rain of Death'.[2]

The 'Fable for Tomorrow' operated as a subverted fairy tale. The early draft began with 'Once upon a time . . .', and promised a romantic story ahead. Carson realised that the best way to implant the notion of impending dystopia was to offer an original Utopia, something to lose. Carson noted in her plans for the chapter, 'Contrast scene of idyllic heavily with nightmare landscape treated with pesticides'. Inspiration for the 'idyllic' came from the classic American image of small town, agrarian living: a rural Midwest town, green, safe and religious (an accompanying picture featured a church steeple). Carson described it as a 'town in the heart of America where all life seemed to live in harmony with its surroundings'. The mention of phrases such as 'heart' and 'harmony' placed the town at the epicentre of the American dream. Farmland merged effortlessly into housing, and humanity seemed in balance with rural nature.[3]

In early versions of *Silent Spring*, Carson gave a name to this town, Green Meadows. The name connoted a place of verdant but tamed nature, a rural realm, Edward Hicks's painting *The Peaceable Kingdom* (1826) rendered in modern American form. Appropriately, 'it lay in the midst of a checkerboard of farms', and sported connections to a traditional agricultural economy. In Green Meadows, Carson captured a postcard of the classic American pastoral idyll. The town reflected Jeffersonian romance with agrarianism. It appealed to the American psyche. It was also very much a place under threat at the time that Carson was writing. Industrialisation, urbanisation, modernity, technology, war, suburbia and now a chemical invasion endangered the survival of the rural locus. In many ways, Green Meadows had already been lost in America, with the fable a lament to the newly departed. On final publication of *Silent Spring*, the name Green Meadows was missing, Carson leaving the ground zero of chemical doomsday open. Scholar Christine Oravec referred to this technique as 'inventional archaeology', an attempt by the author to foster general sympathy for the concept, providing 'an invitation to mythology'. For Carson, the point was that Green Meadows had a 'thousand counterparts in America', and thus needed no moniker.[4]

The fable related the demise of Green Meadows: a town that went silent, where no birds sang. Carson's dystopia was rooted in the breakdown of the cycle of nature by pesticide contamination, of pollination ending, birds dying, and a chain of life being broken. For the reader, the shock came in this discovery. In an early draft, Carson used a visitor returning to town as a guide for the reader, the traveller puzzling what was wrong at Green Meadows, that 'An earthquake or a fire would have wrought disaster he could see and understand. This was something subtle, yet he could sense its presence all around him.' The man found children dying of convulsions and a disturbing silence. The disaster seemed personal to the reader. It also seemed plausible and authentic.[5]

The fable set out the 'dark theme' of *Silent Spring* for its readers. Language proved consciously downbeat. With mention of 'tragedy' and 'disasters' ahead, Carson warned that a 'grim specter has crept upon us unnoticed', chemicals projecting a 'shadow of death' across America. Rather than safe household and agricultural products, DDT canisters were recast as mysterious and deadly. Rather than producing the perfect lawn, they brought a 'strange blight' and 'evil spell' to Green Meadows. Worst of all, 'the people had done it themselves', everyday Americans had invited their dooms-day. Carson's apocalyptic language brought the scale of disaster home.[6]

The nature writer also pondered whether to date this chemical doomsday. One draft read, 'Until the strange and nightmarish events of the year 1965, the community of Green Meadows' was healthy, positing just three years before total disaster. But in notes for Chapter 1, she jotted to herself, 'Better not set specific year. Must cataclysm be so sudden? The end date for the world was deleted from the published version of *Silent Spring*. By shying away from an environmental sell-by date, Carson protected herself from attack. Other environmental writers, such as Paul Ehrlich, took risks by predictions, *The Population Bomb* (1968) derided when its projections of overpopulation and global collapse failed to come true, when his Malthusian-style catastrophe failed to materialise. Like Carson, Ehrlich used the technique of the 'fable/story' as a shock tactic to engage his readers. He depicted a world lost to over-population. With *Silent Spring*, the power of the fable was enough in itself. And evidence of pesticide misuse came chiefly from the past, rather than projected into the future.[7]

The doomsday fable proved crucial to the success of *Silent Spring*. While some claimed the fifty-five pages of notes in *Silent Spring* gave Carson authority over her public, the real power of the book rested in the first few paragraphs. The fable set the tone for the rest of the work. It provided the shock revelation that some-thing widely considered a miracle product for the American home, farm and nation, was actually dangerous to all three. The fable showed what would happen if readers ignored Carson's warnings: the rest of the book provided the evidence and detailed argument, and expanded on the threat of 'biocide'. For scholar Terry Tempest Williams, the storytelling element of *Silent Spring* appealed: 'she weaves together facts and fictions into an environmental tale of life, love, and loss'. As Craig Waddell noted, the 'apocalyptic vision' contributed much to the appeal of the book, while Oravec referred to 'A Fable for Tomorrow' as a 'rhetorical bombshell'.[8]

Caught within 'A Fable for Tomorrow' were all of Carson's key ideas and themes. It provided a message in a bottle bursting with catastrophe culture. It related her criticism of DDT and pesticides, and her worry over science and its accountability. The fable ide-alised nature and highlighted her concern for its fate (and that of humanity, too). It touched on contemporary concerns of the 1950s by referencing nuclear families, Cold War themes, McCarthyist witch-hunts and the nuclear arms race (and fallout worries). The fable celebrated simple, rural America, and the empowered citizen, while remained critical of modern, technological and corporate directions. In one short chapter, Rachel Carson showed how the United States was inviting doomsday in the post-war period.

The chemical world and chemical warfare

The 'Fable for Tomorrow' did not just offer a frightening dystopic vision for America, it also tore down a specific Utopia of the post-war era: the chemical world. The chemical world owed its existence in part to World War II. Tied to the civilian deployment of wartime products (from penicillin to napalm), new pesticides entered the market as miracle cures. Quickly approved by the US Department of Agriculture (USDA), stockpiles of DDT held by DuPont for military use (to protect troops from malaria-carrying insects) entered the domestic front largely untested. By the late 1950s, over 200 'basic chemicals' were on the market. Agriculturalists and home consumers welcomed this flood of pesticides. The sale of DDT connected with naive hopes of a realm of limitless energy (from sources such as nuclear power) and limitless crops (from pesticides). A new chemical world was in tune with a post-war vision of white cities, clean living and material abundance. A rosy American future rested on a new wave of product lines. The promise of a better life was evident in industrial advertising: DuPont, for example, offered 'Better things for better living . . . through chemistry.'[9]

The language of past conflict shaped the chemical offensive. Products of World War II were pitted against a fresh (but traditional) foe in the guise of wild nature. The same arsenal that faced the Nazis was turned against insects. 'The crusade to create a chemically sterile, insect-free world' entailed the projection of nature as dangerous and evil. The chemical industry advertised its wares in such a fashion. Pests became the chief enemy. Threats were overestimated to justify action. The establishment of the chemical world entailed drastic measures to conquer nature, including chemical warfare. Scholar Thomas Dunlap described how for the industry it was a case of 'man and insect locked in battle'. On the frontline of the battlefield, DDT served as a powerful weapon akin to the atomic bomb.[10]

Newspapers covered skirmishes on the front line. In 'Insect War Brooks No Truce', Peter Bart for *The New York Times* reported the 'total war' unfolding on the ground, detailing the $700,000,000 a year spent on consumer-based insecticides, and the hope of a 'remarkable new array of bug-killing weapons' just around the corner. Bart also noted the costs: poisoning of wildlife, contaminated food residues, and insect resistance and immunity. 'History's longest war doubtless is the struggle of man against insect. Despite the length of the battle, however, there is consider-able disagreement over which side is winning,' he concluded. The August 1945 edition of *Time* magazine positioned pesticides as coming straight out of the war, with a new arsenal at the hands of Americans, including the ominously named 'insect bomb' (or aerosol). The wonders of peacetime included the final victory over the buzzing pest: 'It looked as if one of the early blessings of peace would be deliverance from the fly and mosquito. With the Army and Navy releasing some of its new insecticides for civilian use, the war against the winged pests was under way.'[11]

Carson's concern over a chemical world dated back to 1945 when she offered to *Reader's Digest* an article on DDT research at Patuxent Research Refuge in Maryland. Patuxent researched the effects of the pesticide on the environment. Carson wrote to *Digest* explaining, 'Practically at my backdoor here in Maryland, an experiment of more than ordinary interest and importance is going on ... wiping out insect pests ... what other effects DDT may have.' Elmer Higgins at Patuxent alerted Carson to the project. The same year, Edwin Way Teale, an ornithologist, published an article in *Nature* warning of the impact of pesticides. *Time* maga-zine similarly raised concerns. Citing research at the University of California, where attempts to eradicate the codling moth led to the unexpected death of ladybirds, *Time* pondered, 'The more entomologists study DDT, the new wonder insecticide, the more convinced they are that it may be a two-edged sword that harms as well as helps'. *Time*'s verdict: '1) DDT is unquestionably the most promising insecticide ever developed; but 2) it is not yet safe for general use.' The following year, an article in the *New Republic* reported problems with DDT aerial spraying in Pennsylvania, of how 'the sun arose on a forest of silence – the silence of total death. Not a bird broke the ominous quiet.'[12]

A number of incidents caught Carson's attention. From 1952 onwards, Carson had been in correspondence with Olga Huckins, former editor of the *Boston Post*. In 1957, Huckins found a significant number of birds dead outside her home probably due to pesticides. She referred to the incident as a 'harmless shower bath' that quickly turned into mass poison. This spurred Carson to investigate other instances of pesticide misuse. The USDA campaign to eradicate the fire ant proved to be one such case. Situated as a modern insect evil to rival the boll weevil, the fire ant, an exotic to the United States, killed off livestock and bit people. Critics claimed the USDA's tactic with the fire ant was simply to bomb it to oblivion. Ornithologist Robert Cushman Murphy told of 'trigger happy' bombers wreaking havoc. Aerial bombing of fire ants paralleled fire bombing in World War II, only rural areas, not cities being the target. Carson had concerns over the tactic. Also, on the east coast, the USDA launched a campaign to wipe out the gypsy moth. In 1957, spraying over the region of Long Island included cities and suburbs. Residents rallied against the chemical fallout. A court case ensued, lasting three years, with Long Islanders calling on expert witnesses to testify. The conquest of nature from the skies entailed widespread and indiscriminate spraying.[13]

Each of these 'campaigns' against nature featured in the final version of *Silent Spring*. Carson served as correspondent on the front line, reporting on the growing casualties of chemical warfare and the unfolding environmental doomsday. In covering the campaigns, Carson deemed pesticides 'the enemy'. She daubed products such as DDT the 'elixirs of death', linking them with historic poisons, such as arsenic, and modern military concoctions, such as mustard gas (thus underlining a sense of chemical warfare). Carson drew attention to the skull and crossbones insignia of the pesticide world, a portent of death, and classified new insecticides as 'biocides' for the environment.[14]

She also drew on the Frankenstein myth. The chemical industry signified science run amok, the mad, white-coated scientist mixing all kinds of brews with 'fanatical zeal'. They were irresponsible and childish, boys with toys. 'The chemical weed killers are a bright new toy,' wrote Carson. The toys were poorly made and unsafe. Famously, Carson labelled pesticides 'As crude as the cave man's club', and belonging to a 'Stone Age of Science'. The toys also gave a sense of strength. 'They give a giddy sense of power over nature to those who wield them, and as for the long-range and less obvious effects . . .,' she wrote. Pesticides revealed the greater quest of scientists to master and control nature. In early notes, Carson outlined the 'major thesis' of *Silent Spring* to be 'that in at least one major area of man's efforts to gain mastery over nature – the reduction of unwanted or "pest" species – the control operations are themselves dangerously out of control.' She feared a world of 'domination by the engineer'. One newspaper ran the headline, 'Chemical "Frankenstein": Are Pesticides the Monsters that will destroy us?'[15]

Other chemicals that entered the post-war marketplace, ones that had caused significant human harm, became cohorts of DDT. Carson tapped a growing distrust of science and technology thanks to a number of high-profile cases. Thalidomide, horror drug of the period, became DDT's bedfellow. Dr Frances Kelsey of the FDA held back the pregnancy drug after health concerns: she was proven right, with significant child deformities emerging in European countries, where the drug was deemed safe to use. To Carson, pesticides had the same potential to cause genetic problems. She told a *New York Post* representative how, 'It is all of a piece thalidomide and pesticides – they represent our willingness to rush ahead and use something new without knowing what the results are going to be.' In reference to both thalidomide and strontium-90 (a fallout isotope that had seeped into food products), Carson wrote, 'these chemicals are now stored in the bodies of the vast majority of human beings, regardless of age. They occur in the mother's milk, and probably in the tissues of the unborn child.' A chapter titled 'The Human Price' linked modern pesticides with the number-one health concern of the twentieth century: cancer. Carson detailed the cranberry scare of the late 1950s, when a cancer-causing herbicide (aminotriazole) contaminated fruit during the Thanksgiving period, preventing the usual culinary fanfare. The cranberry scare raised concern over how well the USDA regulated its industry, and how dangerous chemicals could be. Highlighting recent panics and scares, *Silent Spring* emerged as a timely reaction to the increasing chemical and technological nature of the post-1945 arena. 'Born of the industrial age', Carson hailed the most dangerous of chemicals as the modern equivalent of smallpox or plague, with little hope of control or survival.[16]

They also spelt the next stage in a persistent narrative of conquest and extinction, a new doomsday stage. As Carson saw it, 'The history of recent centuries has its black passages – the slaughter of the buffalo on the western plains, the massacre of the shorebirds by the market gunners, the near-extermination of the egrets for their plumage', but pesticides represented the latest black passage, 'a new chapter and a new kind of havoc'. By a 'new kind of havoc', Carson meant a path winding towards global collapse, inviting doomsday: 'For the first time in the history of the world, every human being is now subjected to contact with dangerous chemicals, from the moment of conception until death'. Chemicals threatened to 'engulf our environment'. This chemical and technological world spelt a direct challenge to survival. In a sense, Carson was forging her own chemical dystopia, and offering a polar opposite to the white-city Utopia of artificial living. As editor Paul Brooks penned in his forward to *Silent Spring*, Carson warned of 'the poisoning of the earth with chemicals'. She presented a chemical-induced loss of the planet, with the death of nature coming from a 'barrage of poisons'. Such fears spoke to a generation gradually coming to terms with the fragility of Planet Earth.[17]

War against nature

Now in our present time man has acquired unprecedented power to alter the physical nature of his world, in new-found arrogance now extend his goals – far beyond the 'control' of nature to the dream of actual destruction and the substitution of an artificial world. There are those who believe – and I am one of them – that this cannot be done without bringing disaster to man himself. The hour is late, but there is yet time to reconsider our course.

Draft notes for *Silent Spring*[18]

For Carson, the chemical war was not just against insects or pests but against every form of life. It amounted to 'man's war against nature', an invitation to doomsday. Carson held a copy of an address by Roger Heim at the International Union for Conservation in October 1956. Heim declared 'the most total war that man has ever waged, that he henceforward carries on simultaneously at numerous points of the globe . . . is the war which man has declared against Nature on this earth'. In a 1958 letter to Harvard zoologist Dr W. L. Brown, Carson confided that she felt truly disturbed by the 'tampering' with nature through pesticide use, and planned a book that would reveal the 'serious threat to the basic ecology of the earth, and to all its human and non-human inhabitants'.[19]

The 'total war' spelt a paradigm shift in human–nature relations.

Innovations in technology, science, agriculture and industry provoked a fundamental shift in the balance of power. Carson argued, 'Only within the moment of time represented by the present century has one species – man – acquired significant power to alter the nature of his world.' The new arsenal at human disposal, made up of bombs, chemicals and high technology indicated heightened dominance over the environment in the post-war era. Carson also saw the advent of artificial, 'replacement' nature in such work, that, 'The chemicals to which life is asked to make its adjustment are no longer merely the calcium and silica . . . they are the synthetic creations of man's inventive mind, brewed in his laboratories, and having no counterparts in nature'.[20]

Using a language of 'biocide', Carson felt that the latest science and technology posed a clear danger to the country. Total war would lead to total environmental collapse. The chemicals and sprays threatened 'the contamination of air, earth, rivers, and sea with dangerous and even lethal materials . . . This pollution is for the most part irrevocable; the chain of evil it initiates.' She documented instances of salmon accustomed to their annual travel upstream falling victim to a collapsing ecosystem: 'Amid such a picture of death and destruction, the young salmon themselves could hardly have expected to escape, and they did not'. Carson cast nature as innocent and doomed.[21]

These new powers seemed all the worse given the arrogance and ignorance of humanity. On the question of volition and responsibility, the author mused: 'Some of his vaunted "conquest of nature" is deliberate, some is an unintended, often unrecognized byproduct of a deliberate action'. In her notes for *Silent Spring*, Carson scribed the danger of 'anonymous death' corresponding with pesticide use, that it 'eases conscience not to know victims of ones acts . . .'. The current state of civilisation, the human condition, seemed flawed and out of kilter. In book notes, Carson wrote: 'Through some strange confusion or myopia he has come to believe he can poison the environment, an environment he shares with other living things, without suffering the consequences'. Rather than a step towards a more civilised and sophisticated world, Carson suggested pesticides were part of a new culture of killing and a growing lack of awareness of environmental responsibilities. Referring to pesticides as 'caveman's tools', she feared a return to barbarism. 'The question is whether any civilization can wage relentless war on life without destroying itself, and without losing the right to be called civilized,' she mused. To support her contentions, Carson quoted German philosopher Albert Schweitzer: 'Man can hardly even recognize the devils of his own creation' and 'Man has lost the capacity to foresee and to forestall. He will end by destroying the earth.'[22]

Carson's earlier works – *Under the Sea Wind* (1941), *The Sea Around Us* (1951) and *The Edge of the Sea* (1955) – were fundamentally about the beauty of nature. They documented marine life, seashores and oceans. Mostly, they carried a measured but optimistic tone. The books presented Carson as the consummate nature lover and writer, part of a long line of authors celebrating the American landscape, from Henry David Thoreau to John Muir. Only occasionally did Carson's work offer any obvious environmental warning. In *The Sea Around Us*, Carson posed 'The beautiful has vanished and returns not' but mostly the book worshipped the ocean wilderness. Working titles for *Silent Spring* illustrated a novel kind of concern, a fresh battle on the horizon. In discussions with her literary agent, Marie Rodell, Carson considered *The War Against Nature* and *At War with Nature* as options. *Man Against Nature* and *Man Against the Earth* became popular choices for a time. Maine neighbour and confidante Dorothy Freeman referred to *Silent Spring* as 'the poison book' in conversations with her friend. The final title came from a John Keats poem.[23]

Silent Spring thus represented a change of tack, something very different. It was a warning sign, and a clarion call for action. Although very much a Carson book, and, like previous titles, rich in admiration for the natural world, *Silent Spring* was essentially dark in tone. It was not about beauty but about decay, not about life but death. For Carson, this was an important shift of emphasis. It was the result of a traditional, Audubon-card-carrying, wilderness-loving conservationist transforming into a modern environmental thinker, worried by global pollution and modern chemical contaminants. Scholar Mark Lytle saw that 'To protect the natural world that inspired her earlier books, Carson believed she must warn against what she saw as impending disaster.' Biographer Linda Lear emphasised the idea of Carson as dutiful witness: 'she could not stand idly by and say nothing when all that was in jeopardy, when human existence itself was endangered'. Confronted by a modern doomsday scenario, Carson seemed compelled to act. In publishing *Silent Spring*, Carson offered millions of others an easy conduit towards environmental consciousness.[24]

Thirteen years earlier, Aldo Leopold provided a similar rallying cry in *The Sand County Almanac* (1949). The US Forest Service veteran documented the errors in extirpating predators and meddling with wildlife numbers in US quarters, and called for a new land ethic. Leopold had undergone an epiphany in the wilderness, famously witnessing the 'fierce green fire' of a dying mother wolf, and realising the need for humans to respect their environment. Carson similarly lamented how 'By acquiescing in an act that can cause such suffering to a living creature, who among us is not diminished as a human being?' For Carson, her epiphany came not in the wilderness but in the gardens of America, by witnessing the death of the robin and other songbirds from DDT. Carson understood Leopold's wildlife issue, referencing the collapse of deer herds at the Kaibab plateau as a lesson in mismanagement and ill thought-out intervention. Both authors questioned the contemporary battle against nature: for Leopold, predators, for Carson, insects. One war had been simply replaced by another, one pest replaced by another. The victims of the war were discussed in similar ways. Leopold's emotive scene of the wolf mother dying by bullet (it could equally have been strychnine) was paralleled in *Silent Spring* by a variety of poison scenes: the death of the meadowlark, the robin, the salmon, and the squirrel. Carson wrote of the 'mute testimony of the dead ground squirrels', akin to the 'listening' mountains featured in *Sand County Almanac*.[25]

Both Leopold and Carson had scientific backgrounds but wrote with sentiment as well as authority. They saw their role as educators and witnesses. For both authors, the idea of 'going against' the natural order of things concerned them. Twentieth-century intrusions by humanity had upset the 'age-old' patterns of nature. As Carson wrote in a list of basic themes to be covered in *Silent Spring*, 'The sprayers are a negative force working *against* nature, not with it'. Both naturalists believed in ecology, the balance of nature and venerated the 'web of life'. They alike called for the ethical treatment of land and respect for all kinds of life.[26]

Death of nature/bird

The sedge has withered from the lake,
And no birds sing.

La Belle Dame Sans Merci by John Keats

In 'A Fable for Tomorrow', the silence of spring and the evidence that something was wrong were denoted by one factor, the absence of bird song. Like a miner's bird in the cage, still and inert when poisonous gases strike, Carson cast the songbird as nature's monitor, silenced when pesticides spread. The songbird served as a barometer of a healthy environment, and an indicator of crisis.

Songbirds were important to Carson. The death of songbirds in Olga Huckins's garden provided the inspiration for *Silent Spring*. In a letter to the *Boston Herald,* Huckins described the deadly bath experienced by her 'lovely songbirds' in Duxbury, Massachusetts. Carson referred to Huckins's reserve as 'a small world made lifeless'. In the chapter 'Rain of Death', Carson described the death of the robin:

> the once beneficial rain had been changed, through the evil power of the poison introduced into their world, into an agent of destruction. Birds seen drinking and bathing in puddles left by rain a few days after the spraying were inevitably doomed.[27]

Both Huckins and Carson were avid birdwatchers and supported the Audubon Society. The founding statement of the society carried a warning similar to that found in the opening fable of *Silent Spring*: 'We may read [the story] plainly enough in the silent hedges, once vocal with the morning songs of birds and in the deserted fields where once bright plumage flashed in the sunlight'. Carson captured this broad sense of loss in bird numbers, this avian-styled catastrophe culture. The first issue of the *Audubon* magazine in 1887 reprinted an article from the *Spectator*, warning

> There will soon not be a bird of paradise on earth . . . Man will not wait for the cooling of the world to consume everything in it, from teak trees to humming birds, and a century or two hence will find himself perplexed by a planet in which there is nothing except what he makes.

The article was entitled, 'Man the Destroyer'. Carson also raised the spectre of mass bird extinctions. She reported how 'Pigeons are suddenly dropping out of the sky dead' in England, akin to the disappearance of the passenger pigeon in the United States. Citing fears over the demise of the robin and the American eagle, Carson warned, 'We plunge vast segments of our bird population into the night of extinction'. Tightly constructed in the fable, and on display throughout the work: a poignant 'extinction landscape', of silent towns, rivers of death, and avian corpses. *Silent Spring* served as a catalogue of the endangered, a book warning of imminent extinction.[28]

Positioned at the heart of this landscape, the songbird acted as the key symbol of the death of nature. Deadly silence, the end of birdsong signalled the total demise of nature. Instances of bird death recurred throughout the work, a theme of avian tragedy perforating *Silent Spring*. The fable reasserted itself in later chapters, such as 'And No Birds Sing', that related the effects of spraying at East Lansing and the fate of the 'doomed robins'. The silencing of innocent, friendly and near-domesticated creatures had an impact on the audience. Later environmental disasters, such as the Santa Barbara oil spill, carried similar images of birds dying.[29]

Atomic warfare

> Some evil spell had settled on the community: mysterious maladies swept the flocks of chickens; the cattle and sheep sickened and died. Everywhere was a shadow of death.[30]
>
> *Silent Spring*

For most readers in the 1960s, the shadow of death in 'A Fable for Tomorrow' reminded them not just of the perils of pesticides but of another threat in their lives: nuclear Armageddon. Caught up in Carson's fable was a story of a town hit by fallout, a place caught in the grim aftermath of nuclear attack. Unexplained deaths were caused by radiation sickness, the strange stillness accountable by a 'white granular powder' on the ground. An invisible, mysterious force had taken control. The power of the fable rested in its atomic roots. As Ralph Lutts explained, 'It evokes the image of a town dying from nuclear fallout'. Carson reworked this nuclear scene of desolation, something eminently familiar to her readers, and showed how DDT could break down America just like nukes. The similarity of the two dangers was thus brought home.[31]

Released in September 1962, just one month before the Cuban Missile Crisis, *Silent Spring* corresponded with a wave of panic across the United States. With the Cold War at its hottest, nuclear fear skyrocketed. With only a few minutes to midnight on the doomsday clock, missiles being constructed at Cuba, spy planes shot down, naval quarantines ongoing and round-the-clock television coverage of Khrushchev, Kennedy and Castro, danger seemed very close at hand.

Certainly, Carson was well exposed to nuclear fear. As an engaged reader and trained biologist, she kept numerous articles on atomic science. These included a 1955 article in *Newsweek* entitled 'This Dangerous Planet', a prescient and alarmist piece on the environmental dangers of a new atomic era. *Newsweek* asked 'Is the lingering radiation spread around the earth by past or future H-bombs capable of generating another biological catastrophe, an insidious weakening of the human race, perhaps its extinction?' The article surveyed the dangers posed by the atom, reflecting on mutations, genetic threats, and 'dead world' scenarios predicted by current 'prophets of doom' (such as Dr Anderson Nettleship's 'Biopthora' thesis, *pthora* Greek for 'destruction'). Carson also held a copy of *The New Yorker* from January 1955, with 'The Talk of the Town' dedicated to the 'creeping suicide' of bomb-making.[32]

A number of high-profile nuclear incidents also marked the period. The United States military's Castle Bravo test and the poisoning of Japanese trawler *Lucky Dragon* made the news. The discovery in 1953 of strontium-90 in soil, milk and even children's teeth generated further concern. Radioactive milk was no health tonic. The strontium-90 scandal had an impact on a nation caught up in new families. Bomb fears interfaced with the baby boom, irradiated soil and decaying teeth highlighted the spread of fallout beyond the confines of Nevada Test Site (or the Pacific) and into the back yards and bedrooms of America. Meanwhile, in July 1959, *Time* magazine published an article on the need for civil defence and fallout shelters to help cope with the 'major disaster' of nuclear war. *Time* labelled radiation the 'silent killer'. This image of a lurking, invisible threat reappeared in 'A Fable for Tomorrow'.[33]

The nuclear age highlighted fresh kinds of danger to Carson. Lytle claimed that the bomb was fundamental in shaping Carson's views on pesticides, that 'For Carson, a shift in thinking began with the dawn of the atomic age'. Because of nuclear testing, 'she grew ever more uncomfortable with the power of science and technology to undermine or even destroy that ecological interdependence'. The bomb symbolised a modern world out of control, an unstable product of the 'impetuous and heedless pace of man rather than the deliberate pace of nature'. Carson expressed strong reservations over atomic science. In notes for *Silent Spring*, she wrote simply, 'radiation is now the unnatural creation of man's tampering with the atom'. She also confided to friend Dorothy Freeman how, before the bomb

> It was pleasant for me to believe, for example, that much of nature was forever beyond the tampering hand of man – he might level the forests and dam the streams, but the clouds and the rain and the wind were God's.

The bomb fundamentally challenged the 'sanctity of nature'.[34]

Nuclear fear also shaped her writing plans. In a new preface for *The Sea Around Us* (published in 1960), Carson highlighted the threat of nuclear waste to marine life, that radioactive dumping in the sea posed an 'invitation to disaster'. In an early list of pointers for *Silent Spring*, she included the parallel between pesticides and radiation. Other notes read: 'Chemicals – the sinister and little recognized partners of radiation in changing the very nature of the world, the very nature of its life' and that entomology is 'armed with weapons possessing all the deadliness of the atomic age'. As Lutts remarked, 'It is no accident, then, that the first pollutant mentioned by name in *Silent Spring* was not a pesticide, but Strontium 90.' Editor Paul Brooks encouraged Carson to exploit the nuclear climate, instructing 'the parallel of chemicals and effects of radiation is so dramatic that people can't help getting the idea. In a sense, all this publicity about fallout gives you a head start in awakening people to the dangers of chemicals.' For Lutts, nuclear concerns 'pre-educated' Americans, readied them for *Silent Spring*. Unsurprisingly, nuclear references abounded in the final version of the book. Asserting the ties between atoms and pesticides, the first five chapters of *Silent Spring* all began or ended with nuclear imagery. The serialisation of Carson's work in the *New Yorker* also seemed apt. The magazine had published the controversial atomic drama, 'Hiroshima', by John Hershey in 1946, and now came up with a still more dramatic piece.[35]

In part, Carson was right in casting atoms and pesticides as post-war bedfellows. Both were fallout from World War II, products of the modern military–industrial complex, and subsequently marketed as miracle breakthroughs of the post-1945 era, and technological harbingers of a new age . Robert Gottlieb remarked on the 'pesticide explosion' of the period, something to rival nuclear tests. For entomologists, DDT represented 'their atomic bomb', but without all the dangers of fallout. In the home, 'bug bombs' proliferated. Atoms and pesticides were both enrolled in 'Winning the War Against Insects'. The USDA oversaw an on-going war between insect and man, with the modern American farm situated as the key battlefield, the nature–man ground zero. In a 1956 USDA report, W. L. Popham painted a bleak picture of insect devastation. Detailing losses on the scale of $4,000 million a year, Popham bemoaned how 'Insects never tire in their ceaseless battle against man over the inheritance of the earth'. With concerns about 'tomorrow, [where] the cost may be even greater if we wish our high standard of living to continue or improve', Popham predicted that 'Our growing population, our increasing world responsibilities, and our limited and gradually diminishing natural resources force us into very-greater competition with insects for our food, our possessions, and our health'. The great hopes in such a war were science and technology. Noting new insecticides, radiation techniques, and the 'chemical revolution' unleashed by DDT, Popham enthused how 'scientists have come up with some startling new weapons for farmers to use in this elemental struggle for survival'. Aldrin, DDT, radiation – everything possible needed to be thrown at the enemy.[36]

Gathering materials for the writing of *Silent Spring*, Carson collected a number of articles on the specific use of radiation in insect control. As early as 1955, the USDA had announced how 'atomic science gives entomologists new weapons against insect enemies of mankind'. Carson knew of USDA entomologists Edward Knipling and Raymond Bushland who discovered sterilisation through gamma rays as an effective tool in controlling the screwworm fly in the American south-east. Their work drew on theories by Dr H. J. Muller, a geneticist interested in X-rays and mutation, while Oak Ridge laboratories provided the gamma technology. Knipling had great confidence in the project, declaring 'the possibilities for controlling nature's pests by the use of radiation sterilization are limitless'. Along with a report by Knipling, Carson held a copy of a *Saturday Evening Post* article by Charles E. Ball on the sterilisation project. Ball told of the 'major breakthrough' in 'mankind's age-old battle against insects'. 'Scientists are tricking the deadly screwworm fly into breeding itself out of existence,' Ball enthused, detailing a very subtle (almost underhand) form of atomic warfare, the mass sterilisation of males by rays, duly 'released to perpetrate race suicide on their own kind'. The article's headline ran 'Atomic War on Insects'. A USDA leaflet equally portrayed the atom as a 'key weapon' in the fight against insect-driven livestock losses. Titled 'Atoms vs the Screwworm', the leaflet pitted the latest superhero of science and technology against nature's villain.[37]

On a broad, ecological level, Carson aligned pesticides with radioactive isotopes in terms of their potential to cause widespread environmental damage. Carson wrote to Brooks, explaining, 'Whether radiation or chemicals are involved, the basic issue is the contamination of the environment.' Both proved capable of covering vast distances, were hard to detect, lacked research on safe doses, and fell from the sky. Carson dubbed the pesticide threat 'chemical fallout'. On describing the array of new pesticides at America's disposal, Carson relayed how 'Many of the chemical agents in this alarming mélange imitate and augment the harmful effects of radiation'.[38]

Both scientific discoveries posed a threat to humanity. Exploring ties of pesticides with cancer, Carson compared the illness of a Swedish farmer exposed to pesticides to that of the Japanese trawler crew caught in Castle Bravo. Carson found the farmer's poisoning 'strangely reminiscent of that of the Japanese fisherman Kuboyama of the tuna vessel the Lucky Dragon'. Radioactive ash and pesticide dust became indistinguishable, 'For each man a poison drifting out of the sky carried a death sentence'. Carson certainly felt concern over the 'genetic effects of radiation' and the 'same effect of chemicals that we disseminate widely in our environment'. Both threatened to unweave the makeup of the human genome. An early dust jacket for *Silent Spring* read, 'We have employed chemicals with a potential effect on the germ cell and the reproductive process which are closely paralleled by the effects of atomic radiation.' Carson raised the spectre of mass mutation, claiming that, 'many chemicals, like radiation, bring about gene mutations'.[39]

The ultimate spectre was nonetheless the end of the world. Bombs and pesticides threatened total collapse. Both were front-line threats, on the doorstep and in the back yard of America, located in soil samples and garden turf alike. Chemicals and radiation had infiltrated the homeland; together, 'their presence casts a shadow'. Where nuclear war spelt quick death, pesticides represented the kind of slow death captured in 'A Fable for Tomorrow'. While school kids practised 'duck and cover' drills for nukes, no such preparations existed for DDT.

The coming together of these two threats was crucial to the establishment of *Silent Spring* as a novel kind of doomsday text. Mushroom fears combined with fears of nature's collapse. Radiation and pesticides together served as symbols of a new kind of pollution, and a new environmental era. With *Silent Spring*, Carson was 'bridging the gap between the environmental movement and this new fearful vision of Armageddon'. Traditional conservation worries gave rise to the shadow of the bomb, a mushroom cloud rising up behind Yosemite's El Capitan, and a pesticide haze lurking in every American's back yard.[40]

Cold War and the chemical family

No witchcraft, no enemy action has silenced the rebirth of new life in this stricken world

Fable, *Silent Spring*

The extent to which chemical addictions and nuclear fear came together in *Silent Spring* identified it as very much a book (and a doomsday) caught in its period. Written in the late 1950s and early 1960s, *Silent Spring* was very much a Cold War text. In an early draft of the fable, Carson related how the visitor to Green Meadows puzzled over the 'pall of death', looking for an answer through the framework of Cold War logic in 'enemy attack', that 'through some chance combination of atmospheric conditions, there could have been some overwhelming concentration of fallout from the testing of nuclear bombs' while 'wild thoughts of chemical or bacteriological warfare sped through his mind'.[41]

Both chemical agents and atomic energy signified potent weapons of the Cold War. For some industry executives, the insect enemy even resembled the Communist threat. Carson subtly drew attention to this process. She highlighted the 'process of escalation' in chemical products that mirrored the nuclear arms race. The continuing search for 'deadlier' weapons, and how the 'chemical war is never won', directly paralleled the self-perpetuating stalemate of the Cold War. Carson foresaw 'super races immune to the particular insecticide', mirrored the sense of that redundancy and continual replacement in Cold War armaments. With the statement 'No witchcraft, no enemy action has silenced the rebirth of new life in this stricken world', Carson referenced McCarthyism and Soviet threat while implicating pesticides as a commensurate danger. Images of indiscriminate spraying wiping out towns resonated with a readership fearful of missile bombardment by the Soviets. The white picket fences of suburbia were no defence against reds in the bed, or poisons in the garden. [42]

Silent Spring spoke to the culture of the late 1950s and early 1960s. Carson was caught in her moment, the rise of a modern suburban chemical world. The 1950s was both the age of advertising and the age of pesticides. The two came together in the marketplace. All kinds of products entered domestic stores. With the rise of suburbia, special attention focused on the quest for perfect lawns (to match the white picket fences). The manicured lawn of mass-manufactured homes represented an obvious target for products. Advertising included 'Debugging made easy' with Carco 'your garden protector', 'kill pests now' with Scalecide, and 'satisfaction guaranteed' with Borerkil. Bug-geta and End-o-Pest were other products. In one advert 'Ideas from Hudson Live Outside and Love It!' the suburban housewife gazed at her wonderful suburban garden. One newspaper commented how 'Packaged Poisons Sold Like Soap'.[43]

Carson feared for suburbia. Pesticides posed a fundamental threat to the back garden. Nuclear war, bombs in the back yard, presented one danger. But pesticides were a permanent domestic problem, an inescapable reality. Carson investigated the scale of hazards in the modern home. She held a copy of a 1959 public health report detailing a survey of homeowners carried out by the Chemical Specialities Manufacturers Association. Worryingly, only between 8 and 15 per cent of respondents read labels on household insecticides. On the issue of the suburban lawn, Carson warned against chemical crabgrass killers: 'Instead of treating the basic condition, suburbanites – advised by nurserymen who in turn have been advised by the chemical manufacturers – continue to apply truly astonishing amounts of crabgrass killers to their lawns each year'. Carson pondered, 'How lethal these lawns may be for human beings is unknown.' On a broader level, Carson worried over the widespread use of household poisons, fearing their seepage into blankets and clothes.[44]

In an article for *National Parks* magazine, Carson focused on this hard sell of chemicals in suburbia. 'Use of poisons in the kitchen is made both attractive and easy', she declared, highlighting how, 'with push-button ease, one may send a fog of dieldrin into the most inaccessible nooks and crannies of cabinets, corners, and baseboards'. Outside the home, the problem worsened. 'Gardening is now firmly linked with the super poisons', Carson wrote, drawing attention to the quest for the perfect lawn. With the combination of insecticide and power mower, Carson feared how 'the probably unsuspecting suburbanite . . . . [is] raising the level of air pollution above his own grounds to something few cities could equal'.[45]

To shock the suburban middle class out of their chemical comfort zones, *Silent Spring* had to speak to it. Carson spoke of the 'vogue for poisons' of 1950s society and 'the fad of gardening by poisons'. Again targeting crabgrass killers, 'The mores of suburbia now dictate that crabgrass must go at whatever cost', she railed. Carson recognised the role of advertising in allying pesticides with the suburban dream. The 'typical illustration portrays a happy family scene, father and son smilingly preparing to apply the chemical to the lawn, small children tumbling over the grass with a dog', Carson wrote. Such images popularised the iconic 'chemical family' of the period. The advertising industry had made chemical bags on suburban lawns a 'status symbol' and pesticides a social norm. On a broader level, Carson warned about embracing the new speed of life and the marketplace, of taking the wrong road, that might seem 'deceptively easy, a smooth superhighway on which we profess with great speed, but at the end lies disaster'. She drew attention to the case of Long Island and the spraying of vast residential tracts in the hope of eliminating the gypsy moth. Keenly, Carson pointed out that the moth was very much a 'forest insect' and not a suburban dweller. Blaming the ineptitude of the Department of Agriculture at both state and national levels, Carson told how 'They sprayed the quarter-acre lots of suburbia, drenching a housewife making a desperate effort to cover her garden before the roaring plane reached her, and showering insecticide over children at play and commuters at railway stations.'[46]

Carson also manufactured specific fictive doomsdays to highlight the danger. The fable's deathly town was one. The spectre of Green Meadows manifested over new communities, a rain of chemical and radioactive fallout threatening the suburban home. Further into *Silent Spring*, illustrators the Darlings provided an artistic impression of suburbia under attack from the skies by a pesticide 'bomber', its flume of poisons leaving a white, blank space on the page, a vacuum, literally wiping out houses, gardens and life. Those suburbanites that remained went on with their business, seemingly oblivious to the pesticide threat. One continued to spray his own garden, another mowing his lawn. The title of the chapter, 'Indiscriminately from the Skies', suggested little hope of escape. Pesticides invaded suburbia, poisoning the nuclear family like a radioactive storm.[47]

With suburban life under attack, Carson encouraged a new kind of civil defence. Carson sought to empower her readers, pushing the 'right to know' and citizen action. Like a 1950s painkiller numbing the pain, Carson felt the public as a whole was being 'fed little tranquilizing pills of half truth', that they were being duped. In a list of themes for the book, point 4 read: 'the Hitler Technique of the Big Lie', with the 'same assumptions of futile gullibility that dictated TV quiz shows'. Propaganda big and small needed questioning. She was sceptical of the dollar-driven chemical industry, lamenting how the 'public has been sold an idea these things are necessary for "progress". Close examination always reveals it is the making of a "fast buck" that is the motive.' *Silent Spring* attacked 1950s corporate America, 'an era dominated by industry, in which the right to make a dollar at whatever cost is seldom challenged'. Carson jotted down the simple equation in her notes: 'Byproduct of man's pursuit of power and money = contamination of his environment'.[48]

Her text also fuelled a growing middle-class interest in the environment. Carson's books affirmed the wonder of nature at a time when Americans were taking to national parks, the great outdoors, in record numbers. An escape from Cold War anxieties, nature served its historic role as social tonic, a reliever of work pressures and Armageddon frames of mind. *Silent Spring* equally complemented new environmental concerns: wilderness threats in parks, smog in Los Angeles, water pollution at Lake Erie – an America of modern pollution and contaminants.

Ralph Lutts felt that Carson had very much a 'broad constituency' with *Silent Spring*. She mobilised everyone 'from bird watchers, to wildlife managers and public health professionals, to suburban homeowners' in tackling the 'environmental threat'. Lear agreed. Carson shocked nuclear families into action: 'She wrote a revolutionary book in terms that were acceptable to a middle class emerging from the lethargy of postwar affluence and woke them to their neglected responsibilities.' Interestingly, as an unmarried woman with a personal life far removed from the nuclear family idyll, Carson herself hardly fitted social norms of the period. As Lytle noted, 'Certainly Carson's life hardly conformed to then current conventions of gender and family.' She criticised the establishment and provided her own dystopia to contrast with the 1950s dream. Certainly, *Silent Spring* bore an element of Betty Friedan's *The Feminine Mystique* (1963) in challenging suburban conformity. But this time, males in white coats (rather than grey flannel suits) represented the enemy, and the silencers of the female voice.[49]

The response

The publication of *Silent Spring* met with outrage from the chemical industry. In response to the serialisation in *New Yorker* magazine, Velsicol, manufacturer of heptachlor and chlordane, threatened legal action. Velsicol's General Counsel Louis McLean proved outspoken in his criticism of Carson. The conflict over pesticides became very much a public-relations war.

The chemical and agricultural industries mobilised a wide range of resources against *Silent Spring*. The National Agricultural Chemical Association (NACA) spent $250,000 on pamphlets, advertising and public relations during the period. The industry targeted the author as much as the publication. The Agricultural Research Service presented Carson as a Thoreauvian nature lover, out of kilter with modern society:

> The balance of nature is a wonderful thing for people who sit back and write books or want to go out to Walden Pond and live as Thoreau did. But I don't know of a housewife today who will buy the type of wormy apples we had before pesticides.

Dr Robert White-Stevens from American Cyanamid proved vitriolic in his criticisms of Carson, and labelled her 'a fanatic defender of the cult of the balance of nature'. Personal attacks on Carson included questions over her scientific prowess, and accusations that her data were very much flawed. The NACA published a 'Fact and Fancy: A Reference Checklist for Evaluating Information about Pesticides' that critiqued extracts from the book. Gender also factored into assessments. Carson was dubbed sentimental, even hysterical. Former agriculture secretary Ezra Taft Benson pondered why 'a spinster with no children is worried about genetics'. The reaction seemed all the more significant given that Carson was taking on a 1950s power elite, a male-dominated industry.[50]

Industry defenders were most horrified by her attack on chemical America. For people like McLean and White-Stevens, *Silent Spring* threatened the chemical-fuelled progress of a nation. On a *CBS Reports* feature on *Silent Spring*, White-Stevens, as spokesperson for the chemical industry, warned, 'If man were to faithfully follow the teachings of Miss Carson, we would return to the Dark Ages, and the insects and diseases and vermin would once again inherit the earth'. A ban on pesticides spelt an almost biblical level of damnation. For William Darby, biochemist at Vanderbilt, Carson's vision threatened 'the end of all human progress'. Exploiting Cold War fears, Velsicol's McLean warned how, by following Carson's advice on reducing pesticides, 'our supply of food will be reduced to east-curtain parity'. Carson was even labelled a Communist.[51]

One response was to combat Carson's dystopia with an alternative dystopia: that of a chemical-less world. Several organisations fashioned doomsday scenarios of poverty, famine and collapse to rival 'A Fable for Tomorrow'. The October 1962 edition of Monsanto's postal magazine featured 'The Desolate Year', an unsubtle horror story of a world without pesticides, and a literary parallel to Carson's fable. Monsanto described the fall of the American agricultural community, detailing how 'the garrotte of Nature rampant began to tighten' without chemical protection, and very quickly 'The bugs were everywhere. Unseen. Unheard. Unbelievably universal.' An invasion of mosquitoes, cattle grubs, boll weevils, other insects, and a plague of grasshoppers terrorised farms. Flesh-eating ticks 'leaped onto people. While the sharp and grasping pincers held fast, the razor-like cutting tools sliced deeper and deeper into the flesh.' The scene resembled a nature-based horror film (like *Them!* or *The Swarm*), with Monsanto telling how victims succumbed to the 'fiendish torture of chills and fever and the hellish pain of the world's greatest scourge'. Aimed squarely at Carson's fable, Monsanto stated that, 'The terrible thing about the "desolate year" is this: Its events are not built of fantasy. They are true.' In *Chemical Week*, biologist (and Sierra Club devotee),Thomas H. Jukes from American Cyanamid provided his version of the fable, 'A Town in Harmony'. Jukes described a town in 'biological balance in Nature' being rife with disease, starvation and famine: 'As the sun went down, the buzzing of mosquitoes could be heard in the town' and 'the last slanting rays of the sun lingered on the small headstones in the town graveyards. Here slept the children who had perished from diphtheria, scarlet fever and whooping cough.' Jukes described a scene of abject horror, a town besieged by rats, insects and disease. Such chemical nightmares disguised some very real and practical fears of the industry: over-regulation, economic curtailment, new testing and ultimately a loss of business (and profit).[52]

Some news agencies sided with the chemical industry. The *Richmond News Leader* published 'The Desolate Year'. R. W. Bales, writing for the *Augusta Herald*, railed:

> to the casual reader who has survived a couple of wars, who even now is caught up in a world crisis, who has contemplated the dangers of fallout and the horrors of nuclear destruction, the case against pesticides seems unnecessarily exaggerated.

DDT appeared a minor concern compared with global conflict. Reflecting the conservatism of the period, a frustrated Bales quizzed, 'surely there is some sane middle ground ignored in *Silent Spring*?'. Another journalist moaned how Carson 'stressed only the dark side'. *Better Homes and Gardens* magazine warned people not to be 'panicked into throwing out all dusts and sprays', as the American lawn needed its chemicals to survive.[53]

A far wider range of media supported Carson, however. Not only the *New Yorker* serialised *Silent Spring*, daily newspapers reprinted whole sections. Much of the media adopted Carson's lines of enquiry. The nuclear angle emerged as one popular reference point. One newspaper carried the headline 'Deadliest Fallout', provoking some confusion over the chemical or atomic threat on the horizon. After relating the danger of radioactive fallout, the *Morning World Herald* warned, 'there may be a greater and more imminent pollution danger in our country today. Unlike nuclear radiation, it can be stopped without the co-operation of Nikita Kruschev [sic].' Virginia Kraft, writing for *Sports Illustrated*, and in support of the chemical industry, felt that *Silent Spring* 'made Nevil Shute's *On the Beach* seem almost euphoric by comparison', and referred to pesticides as 'the life giving spray'. Highlighting cancer fears and genetic qualms, reporters mostly focused on the human angle. 'Miss Carson fears man can destroy himself through indiscriminate use of chemical pesticides,' reported one Cleveland paper. 'Is Today's War on Insects a Peril to Man's Tomorrow?' read another. One headline read: 'Man to Join Dinosaur: Step by Step to Death'.[54]

A sense of alarm swept state and national press. Newspapers, magazines and television fuelled the controversy. One Pittsburgh paper described the unfolding 'Noisy Autumn'. For scholar Priscilla Coit Murphy, 'the media saw merit in playing a continuing, catalyzing role in the debate'. *CBS Reports* invited Carson for interview. Given the risky nature of the topic, corporations responded by pulling their ads.. Roy Attaway for the *Charleston News Courier* related, 'I am not an alarmist. I have no desire to create panic. But I am concerned.' Attaway feared the chemical project unfolding, questioning, 'progress toward what? Slow annihilation?' 'Are we further destroying that precarious membrane that has bound together earth's life lo, these many millennia?' he mused. The *San Francisco Examiner* carried the headline 'Will Race of Man Destroy Itself', explaining the process by which 'man has forgotten he is a child of nature – Or God – [and] has come to believe that he can "conquer" nature, and in his attempts is very likely to destroy his own place in nature and thus himself'. The *Examiner* warned how 'this generation's manipulation of the basic elements of nature, in drugs, in pesticides, in radiation, is not under control'. For the *Examiner*, 'dire prophecies were once the exclusive province of science fiction: but we should have learned from the realm of space exploration that what is fiction today is fact tomorrow'. Environmental doomsday seemed truly on the horizon. Headlines included: 'Everything Began to Change – Everywhere was a Shadow of Death', 'A Grim Spectre', and 'Death Around Us'. The August 1962 edition of *Newsweek* headlined 'Kiss of Doom?' The press did much to highlight the unfolding doomsday scenario.[55]

They also did much to promote the significance of Carson's text. The *Albuquerque Tribune* dubbed *Silent Spring* 'the horror story of the decade', a bold claim, given eight years remained. In the reviews section of the *Berkshire Eagle*, C. Roy Boutard likened *Silent Spring* to Upton Sinclair's *The Jungle*, poisonous chemicals analogous to dirty meat-packing. Equally, Carson resembled social reformer Jacob Riis, who documented the tenements and immigration maladies of New York circa 1900 in calling for public education and action. All three authors highlighted the horrors of the modern industrial world. Others compared *Silent Spring* with Harriet Beecher Stowe's *Uncle Tom's Cabin* (1852) for its capacity to refashion public opinion. Brooks himself suspected, 'Perhaps not since the classic controversy over Charles Darwin's *The Origin of Species* just over a century earlier had a single book been more bitterly attacked by those who felt their interests threatened.' The chemical industry symbolised the new church of the mid-twentieth century, preaching the heady gospel of scientific progress, with Carson cast as dissident. Carson ranked alongside a large group of environmental and social reformers who took to the text to state their case. In contemporary circles, Carson's *Silent Spring* sat alongside Betty Friedan's *The Feminine Mystique* and John Kenneth Galbraith's *The Affluent Society* (1958) in currency and controversy. It presaged a period of criticism focused on industry and government, and showed the power of the exposé years before Watergate and the Tet Offensive.[56]

Thanks to its interface with the period and the power of the doomsday fable, *Silent Spring* led to a range of changes in America. The Life Sciences Panel of the President's Science Advisory Committee (PSAC) report vindicated most of Carson's concerns, and led to a ban on DDT by 1972. The PSAC report stated that, 'until the publication of Silent Spring, people were generally unaware of the toxicity of pesticides'.[57]

Silent Spring was also credited with ushering in environmental consciousness. Senator Ernest Gruening of Alaska labelled *Silent Spring* the *Uncle Tom's Cabin* of the environment. Carson's work highlighted a broader outlook on conservation, shifting activism away from wilderness issues and towards bigger fears over the loss of life by modern contaminants, including radioactivity and pesticides. It was an authentically modern work, no longer tackling industrialisation or resource extraction but documenting chemical attack on American soil, and silent and invisible death. In preparatory notes for the book, Carson wrote, 'The pesticide problem cannot be understood except in context. It must be understood as part of the whole which it is a fragment, the pollution of our total environment.' In December 1963, Carson received the Audubon Medal for achievements, her last public appearance before dying from cancer. For many environmental activists, *Silent Spring* was their personal beginning, their introduction to ecological issues.

The same year as the first Earth Day 1970, Joni Mitchell penned 'Big Yellow Taxi' (1970), an environmental ballad. Mitchell dedicated one stanza to DDT, reflecting the pivotal position of pesticides to modern environmental consciousness. 'Hey farmer farmer, Put away that DDT now, Give me spots on my apples, But leave me the birds and the bees, Please!, They paved paradise and put up a parking lot,' Mitchell sang. On writing the preface to the twenty-fifth anniversary edition of *Silent Spring*, Al Gore reflected on Carson's 'cry in the wilderness', and the immense changes that followed, 'powerful proof of the difference that one individual can make'. The true significance of *Silent Spring*, however, was in mapping out the trajectory of the United States as it hurtled towards doomsday. Carson showed how the American chemical industry had started out on a path of environmental mortification in the post-war period. *Silent Spring* was the wake-up call.[58]

Aside from the banning of DDT, in the end few listened to Carson's broader concerns over the embracing of a chemical nation. Half a century on, the threat of *Silent Spring* still lurked in the back yards and fruit fields of twenty-first-century America. Theories abounded over the collapse of bee colonies. Some blamed 'Colony Collapse Disorder' on clothian, a newly marketed pesticide. Meanwhile, Cornell Laboratory of Ornithology estimated around seven million bird deaths each year due to pesticides. Twenty-first-century America seemed to have more pests and more pesticides. The problem of insect resistance persisted while EPA figures showed around $12 billion dollars spent each year on pesticides in the United States in 2006 and 2007, amounting to a third of the world's expenditure on such chemicals.

Outside the United States, DDT continued to be exported post-1972. In 1969, Union Carbide opened a factory in Bhopal, India. In 1984, a chemical accident there led to the release of methyl isocyanate gas, and 3,787 deaths. The post-*Silent Spring* world included chemical dumping at Love Canal, near Niagara Falls, that led to mass evacuations in the late 1970s and the Philips Disaster, a chemical accident at Pasa, Texas, in 1989. Issues of corporate and scientific accountability lingered, with legal clerk Erin Brockovich discovering contaminated drinking water in Hinkley, California in 1993. If anything, the toxic and chemical world grew rather than shrank in the aftermath of *Silent Spring*.

The advent of genetically modified crops promised some reduction in the need for pesticides. GM crops could be 'engineered' to fight pests, to win the war against nature. But genetic tinkering revealed a new level to the control of nature and the remaking of agriculture. Equally, the quest for the perfect suburban lawn continued, with companies such as Funk Lawn Care from New York 'Growing a better environment'. If not, artificial nature in the guise of Astroturf proved another option. And apples continued to be sprayed, at the top of the list of most contaminated produce in an Environmental Working Group investigation for 2011.[59]

Notes

1. Priscilla Coit Murphy, *What a Book Can Do: The Publication and Reception of Silent Spring* (Amherst: University of Massachusetts Press, 2005), p. 77.
2. Rachel Carson, 'Man Against the Earth', plan 38/680 YCAL MSS46 Beinecke Rare Book and Manuscript Library, Yale (herein all numeric references from Beinecke).
3. Carson, 'plans' 50/900; Carson, 'Opening chapter ideas' 50/899; Rachel Carson, *Silent Spring* (Boston: Houghton Mifflin, 1962), p. 1.
4. Carson, 'plans' 50/900; Christine Oravec, 'An Inventional Archeology of A Fable for Tomorrow', Craig Waddell, ed., *And No Birds Sing: Rhetorical Analyses of Rachel Carson's Silent Spring* (Carbondale: Southern Illinois University Press, 2000) pp. 43, 56; Carson (1962), p. 3.
5. Carson 50/900.
6. Murphy (2005), p. 53; Carson 50/900 and Carson (1962), pp. 1–3.
7. Carson, chapter 1, 50/900; chapter 1, 50/899.
8. Terry Tempest Williams, 'One Patriot' in Lisa Sideris et al., *Rachel Carson: Legacy and Challenge* (Albany: State University of New York Press, 2008), p. 18; Waddell (2000), p. 9; Oravec (2000), p. 43.
9. Carson (1962), p. 7; Mark Hamilton Lytle, *The Gentle Subversive: Rachel Carson, Silent Spring and the Rise of the Environmental Movement* (New York: Oxford University Press, 2007), p. 139.
10. Carson (1962), p. 12; Thomas R. Dunlap, *DDT: Scientists, Citizens, and Public Policy* (Princeton: Princeton University Press, 1981), p. 14.
11. Peter Bart, 'Insect War Brooks No Truce', *New York Times*, 6 May 1961, 38/680; *Time*, 27 August 1945.
12. Linda Lear, *Rachel Carson: Witness for Nature* (New York: Holt, 1998), pp. 118–19; 'Science: DDT Dangers', *Time*, 16 April 1945; John Terres, 'Dynamite in DDT', *New Republic*, 25 March 1946.
13. Lytle (2007), p. 123.
14. Carson (1962), pp. 15, 8.
15. Carson (1962), pp. 12, 68, 297; Carson 50/899; 'Chemical Frankenstein', *Jamaica Press* 60/1072.
16. Lytle (2007), p. 145; Lear (1998), p. 412; Carson (1962), pp. 16, 187.
17. Carson (1962), pp. 85, 15, 187, xiii.
18. Carson 50/899.
19. Carson (1962), p. 7; Heim address, October 1956, 38/680; Carson to Dr W. L. Brown, Museum of Comparative Zoology, Harvard, 5 October 1958, 42/767.
20. Carson (1962), pp. 5, 7.
21. Carson (1962), pp. 6, 131.
22. Carson notes, 38/680; Carson (1962), p. 99; Schweitzer: 38/680 and Murphy (2005), p. 31.
23. Lytle (2007), p. 61. On title, see Lytle pp. 2, 156, Carson 38/680.
24. Lytle (2007), p. 7; Lear (1998), p. 4.

25. Carson (1962), pp. 99–100; Aldo Leopold, *A Sand County Almanac* (New York: Oxford University Press, 1949).
26. Carson 38/680; Carson (1962), p. 64.
27. Carson (1962), pp. ix, 92.
28. *Audubon* magazine 1/1 (February 1887); Carson (1962), pp. 124, 105.
29. Carson (1962), p. 107.
30. Carson (1962), p. 2.
31. Carson (1962), p. 3; Ralph Lutts, 'Chemical fallout: Rachel Carson's Silent Spring, radioactive fallout and the environmental movement', *Environmental Review* 9/3(1985), p. 222.
32. 'This Dangerous Planet', *Newsweek*, 17 January 1955 and *The New Yorker* (January 1955), in Carson 36/563.
33. 'Civil Defense: Against the Silent Killer', *Time*, July 1959.
34. Lytle (2007), p. 120; Carson (1962), p. 7; Carson 38/60; Lear (1998), p. 220.
35. Lear (1998), pp. 373, 375; Carson 50/898, 38/680; Carson (1962), p. 6; Lutts (1985), pp. 212, 221; Murphy (2005), p. 56.
36. Robert Gottlieb, *Forcing the Spring: The Transformation of the American Environmental Movement* (Washington, DC: Island Press, 1993), p. 82; W. L. Popham, 'Winning the War Against Insects', USDA (December 1956), 30/535.
37. USDA Press Release (December/January 1955), Knipling, 'Control of S-W Fly by Atomic Radiation', Charles E. Ball, 'Atomic War on Insects', *Saturday Evening Post*, 9 September 1961, 'Atoms vs Screwworm, USDA leaflet (January 1958), all in 30/534 and 535.
38. Lear (1998), p. 366; Carson (1962), pp. 156, 39.
39. Carson (1962), pp. 229, 229–30, 37, 8; Carson 50/898.
40. Carson (1962), p. 188; Lutts (1985), p. 223.
41. Carson, early drafts, chapter 1, p. 7, 50/900.
42. Carson (1962), p. 8.
43. See www.growingwithplants.com/2010/05/vintage-pesticide-advertising-round-up.html.
44. Carson (1962), pp. 80–1; Rachel Carson, 'Beyond the Dreams of the Borgias', *National Parks* magazine (October 1962), 60/1066.
45. Ibid.
46. Carson (1962), pp. 297, 177–8, 277, 158.
47. Carson (1962), pp. 154–5.
48. Carson (1962), p. 13; 'Basic themes' and notes, 38/680.
49. Lutts (1985), p. 211; Lear (1998), pp. 4–5; Lytle (2007), p. 135.
50. Lear (1998), p. 413; Lytle (2007), p. 178; Murphy (2005), pp. 79, 106.
51. Lytle (2007), pp. 183, 174, 165.
52. 'The Desolate Year', *Monsanto magazine* (October 1962); Thomas Jukes, 'A Town in Harmony', *Chemical Week*, 68/1216.
53. R. W. Bales, 'Isn't it exaggerated', *Augusta Herald*, 20 November 1962, 61/1078; 'Chemists of Pesticide Firms Dispute Author', Cedar Rapids *Gazette*, 14 September 1962, 61/1077; 'American Lawn Under Attack', *Better Homes and Gardens* (November 1962); Murphy (2005), p. 146.
54. 'Deadliest Fallout', Beaver-Falls *News-Tribune*, 3 October 1962 68/1208; 'Chemical Pollution', *Morning World Herald*, 22 July 1962, 69/1209; Kraft: Murphy (2005), p. 145; 'Pesticides Killing Wildlife', Cleveland *Plain Dealer*, 61/1077; 'Is Today . . .', Washington *Star* 25 November 1962, 60/1070; 'Man to Join Dinosaur', *Atlanta Journal* (28 November 1962) 60/1072.
55. 'Noisy Autumn', *Post-Gazette and Sun Telegraph* (Pittsburgh), 2 October 1962, 68/1208; Murphy (2005), p. 129; Roy Attaway, 'An Unpleasant Thought', *News Courier* (Charleston), 28 October 1962, 61/1076; 'Will Race of Man Destroy Itself', *San Francisco Examiner* 22 September 1962, 68/1208; 'Kiss of Doom?', *Newsweek*, 6 August 1962, 68/1208.
56. 'The Tribune Bookshelf', *Albuquerque Tribune*, 22 December 1962, 61/1078; C. Roy

Bourtard in *Berkshire Eagle*, 29 September 1962, 61/1078; Paul Brooks, *The House of Life: Rachel Carson at Work* (1972), p. 293.

57. Lytle (2007), p. 184.
58. Sideris (2008), p. 22; Carson 38/680; Gore in *Silent Spring* (1997 [1962]) Introduction.
59. See EWG Press Release, 13 June 2011: www.ewg.org/foodnews/press/.

CHAPTER FIVE

Black Days: The Santa Barbara Oil Spill and *Deepwater Horizon*

'There's a whole ocean of oil under our feet! No one can get at it except me!'

Plainview, *There Will Be Blood* (2009)

In *There Will Be Blood* (2009), Plainview (Daniel Day Lewis) is an oil prospector in early twentieth-century California with a career obsession. Nothing stems the flow of oil. The black stuff is his life, his blood. It also destroys everything in its path. The film resonates with broader American attitudes to oil. Throughout the twentieth century, the black stuff fuelled the rise of the United States. An energy landscape based around oil determined forms of mass transport, city design (as in Los Angeles), national energy strategy, foreign policy decisions, and even war. American addiction to crude energy amounted to a national obsession. Oil dependency also spawned continual fear of loss. Popularly termed the 'lifeblood of civilization', a future without oil connoted a world of barbarism and dereliction in the popular imagination. Films, such as the *Mad Max* series (1979–85) whereby marauders battle for the last remaining cans of gasoline, captured such anxiety, doomsday landscapes tied to the disappearance of fuel. Yet the presence of oil produced its own discrete set of problems. In environmental terms, a century of oil dependency contributed to global warming, acid rain, and carbon dioxide poisoning. The United States experienced a series of environmental disasters, black doomsday landscapes, in the guise of oil spills at Santa Barbara (1969), *Exxon Valdez* (1989), and *Deepwater Horizon* (2010). This chapter explores the first large-scale disaster, the Santa Barbara oil spill, and the most recent, *Deepwater Horizon*.

The Santa Barbara spill

The use of oil in the energy landscape is nothing new. In California, Native Americans utilised tar from ocean seeps for the manufacture of weapons, utensils and boats. Establishing missions along the California coast in the late eighteenth century, Spanish Franciscans catalogued a range of energy resources as part of their colonisation project. In 1776, Father Pedro Font spotted tar seeps on the shores of present-day Goleta, close to Santa Barbara. The oil revolution began in earnest during the latter stages of the nineteenth century in the eastern states. In 1859, at Titusville, Pennsylvania, Colonel Edwin L. Drake hit oil. His 70-foot (21-metre) well produced twenty-five barrels a day, and was the first successful drilling in the United States. Similar to the effect of mineral rushes on settlements in the West, Titusville became a boom town overnight. Pipelines, steam locomotives and money flowed in and out of town. People converged on the fledgling township. The local population spiralled from 250 to over 10,000. President Ulysses Grant visited Titusville in 1871, imparting his thanks for the export of oil during the Civil War: 'I cannot help remembering what this section of our common country did during the war. I remember the men you sent to the front, and more than that, the value of the staple product of this nation.' A large-scale oil industry emerged in the decades that followed, the rise of the United States tied to its exploitation of the thick black stuff. In the American West, production began in 1865 in Humboldt County, the Los Angeles region from the 1880s and at Spindletop, Texas in 1901. In his novel *Oil!* (1927), left-leaning author Upton Sinclair explored the shenanigans of oil barons and drill workers in southern California in the 1920s. Sinclair painted a picture of power, politics and corruption. Oil had a darker side. Inspired by *Oil!*, *There Will Be Blood* (2009) highlighted the personal demons and horrors of oil development through the eyes of oilman Plainview.[1]

In 1896, Summerland, the first offshore United States oil field, opened off the coast of Santa Barbara. Alongside Texas and Alaska, the region emerged as one of the key oil producing areas for the United States. In February 1968, the federal government leased 500,000 acres (197,000 hectares) of the Santa Barbara Channel to several oil corporations including Philips and Continental in exchange for 603 million dollars. The companies immediately spent $10.5 million on twelve drilling wells. The deal came at a critical time for the American oil industry. With hope of finding new oil, a gold-rush mentality infused the industry, a boosterist mentality to rival the embracing of nuclear power and uranium ore a decade earlier. But later that same year, *Oil and Gas Journal* reported, 'disappointment over initial results', that 'after almost six months of drilling, the Santa Barbara drilling channel has lost some of its luster'. *Forbes* magazine called it 'Black Gold Blues'. Union Oil erected its first offshore platform in September 1968. At position 402, 5 miles (8 kilometres) offshore, Union began drilling at Platform A on 14 January 1969. In 188 feet (57 m) of water, and 3,500 feet (1,067 m) down, the rig struck oil. On 28 January, however, a blowout occurred at the platform. The drill ruptured, and significant oil and gas leaks followed. Fissures in the ocean floor emerged, and black stuff flowed out in uncontrollable fashion.[2]

In response to the unfolding disaster, Union Oil attempted to plug the drill hole and seal fissures in the ocean bed. Over the next ten days, more than 3 million gallons (80,000 barrels) of crude oil flowed. Officials declared a state of emergency. The spill made national headlines. Reports came in of a 'Huge Slick off California Spreads'. The *LA Times* detailed a 'vast and steadily growing' slick engulfing the coastline, of marine wildlife trapped and imperilled, and workmen desperately trying to 'kill the well'. The Santa Barbara oil spill was labelled a huge environmental disaster. California congressman Charles Teague called it the 'greatest catastrophe since the San Francisco Fire'.[3]

Union Oil personnel, US Fish and Wildlife experts, conservation charities and local volunteers rushed to the shorelines. The clean-up arsenal included log and plastic booms and straw for beaches to soak up the oil. Operation Sea Sweep enlisted tugs and barges to 'confine' the slick. The press detailed the huge fight for the environment, of nature, humankind and technology pitted against oil. The *San Francisco Chronicle* reported a 'Battle off the coast', citing talcum powder as the chief weapon against oil. Another headline read, 'Desperate fight to save harbor from Sea of Oil', the slick

Figure 5.1 President Richard Nixon visits Santa Barbara following the oil spill (courtesy of Nixon Library).

'wallowed menacingly' just half a mile (800 m) from the harbour, a plastic boom the only line of defence. People fretted over where the uncontrollable tide might take the black stuff. The *New York Times* reported, 'Life in Santa Barbara today is somewhat reminiscent of civilian life in a war-zone'.[4]

The application of chemical agents in the clean-up furthered the environmental controversy. The use of dispersants, usually employed to clean the engine rooms of ships, wreaked havoc off Britain's Cornish coast when applied to the marine environment during the *Torrey Canyon* spill of 1967. The deployment of 10,000 tons of toxic chemicals caused more damage than the spill itself. At Santa Barbara, the *Sierra Club Bulletin* derided chemical salesmen converging on the seaside town to sell their wares during the emergency. The Guardian Chemical Company exploited the spill in the mainstream media. In the advertisement 'this is for the birds', Guardian sported the use of Polycomplex A-11 as the 'one bright spot' in disaster. Chemical warfare pitted one resource against another, solvents versus petrochemicals. Aircraft targeted the growing spill, the spraying of beaches and coastal waters with the chemical dispersant COREXIT reminiscent of agriculturalists spraying fields with DDT a few years earlier. While *Silent Spring* depicted birds dying from the spray of the pesticide industry, chemicals now purportedly saved many avian lives. An *LA Times* cartoon attempted a seal's perspective on the increasing chemical degradation of its California environment: 'If the oil doesn't get you, DDT will'. In practice, out of 1,575 creatures taken in at cleaning stations, 1,406 died.[5]

At a governmental level, city and state authorities acted on emergency protocol and established an aid package for Santa Barbara. The federal government halted drilling in the channel, and Congress launched investigations into the accident. State senator Alan Cranston and congressman Charles Teague worked on bills to terminate drilling and suspend oil development along the California coastline. On 21 March, President Richard Nixon visited Santa Barbara. Nixon sympathised with those affected, relating:

> The Santa Barbara incident has frankly touched the conscience of the American people . . . What is involved is the use of our resources of the sea and of the land in a more effective way and with more concern for preserving the beauty and the natural resources that are so important to any kind of society that we want for the future.[6]

At a local level, the spill stunned the affluent beach community of Santa Barbara. An environmental disaster on the doorstep shocked a white middle class accustomed to a clean environment and a high standard of living. The spread of oil posed immediate dangers to beach recreation, tourism, homes and businesses. With the ocean contaminated, vessels from the fishing industry rallied to stem oil rather than catch fish. A dark cloud hovered over the popular resort as the black stuff took control of beaches. Tourists stayed away. The loss in revenue amounted to approximately $1 million over eight months. Oil infiltrated beaches and watersides. The manager of an undersea garden feared oil leaking into his tanks, fantasising, 'It would be like putting our fish in the gas chamber'. Residents felt like they were living in a doomsday landscape. Business magnates moaned how 'eighty percent of our economy is based on our environment'. Santa Barbarans rallied to protest against the spill and 'get oil out'. The Santa Barbara oil spill allied with an image of oil dystopia and a doomed environment. The next sections explore the protest and the disaster.[7]

Protesting against the black stuff: Get Oil Out! (GOO)

Two days after the spill, locals founded the protest group Get Oil Out! (GOO). Formed in the midst of environmental disaster, GOO pulsed with energy. It quickly amassed two thousand members. Protest demands included the shutdown of industry operations in the region, no new oil leases, legislation to curb drilling, and the immediate removal of rigs from the channel. Protest techniques reflected the full panoply of activist resources emblematic of the period. Traditional conservationist lobbying combined with radical action born in the 1960s era of civil rights, peace and feminist crusades. GOO was caught in the convergence of old- and new-style activism.

Familiar in conservation organisations, such as the Sierra Club and the Audubon Society, GOO drew on established staples of organising, including town meetings, petitions, lobbying and letter-writing. On 8 June, two thousand people attended a town meeting to discuss the spill. A petition against oil proved particularly effective, netting 100,000 signatures by 19 May. Letters to the president proved popular: Nixon was reminded of his native Californian status, his obligation to protect the coastline, and his commitment to new environmental initiatives (a spate of environmental legislation followed including a revised Clean Water Act 1970, the establishment of the Environmental Protection Agency, and the Endangered Species Act of 1973). Residents called on the president to show concern about 'an area practically on his own doorstep', especially given his 'recent expressions of concern over environmental pollution'. Appealing to Nixon to intervene, local Thomas Quinn waxed lyrical:

> Mr President, bring your little houseboat out to Santa Barbara Channel, float on a cushion of oil, sniff the oil-scented breezes and suffer with the rest of us. Think about things like greed, man's inhumanity to man; the power of the almighty dollar; and hopelessness. The hopelessness of a people fighting an octopus called 'oil interests'.[8]

Santa Barbarans feared the loss of their scenic coastline, demonstrating an aesthetic concern that harkened back to old-school wilderness valuation. Defending the 'natural beauty' of central

California, GOO demanded the preservation of wild beaches for 'future generations'. Article II of the GOO constitution stated the need for: 'The preservation and conservation of the natural charm and beauty of the Santa Barbara Channel', a statement that echoed the tone of older National Park Service mandates. GOO exhibited a nature/conservation dynamic true to the Sierra Club, Audubon Society and the Wilderness Society. It also reflected a distinctive California concern in the 1960s over the loss of remaining stretches of coastline to industrial and residential encroachment, amounting to a last stand against development in a state known (and struggling) with its dual identity of golden beaches and technological frontiers.[9]

GOO equally drew inspiration from contemporary civil rights, free speech and peace crusades, pioneering a range of protest theatrics new in conservation circles but inveterate to the period. The group organised rallies, fish-ins, blockades, public theatre and citizen education. On 8 February, eight hundred people rallied against Union and co. In July, a flotilla (dubbed a 'friendly protest') of thirty-five to forty boats surrounded Platform A as a protest statement. One GOO shipmate related how, 'This whole thing reminds me of the Commanches circling the wagon trains, and history tells us the wagon trains won'. Santa Barbarans signified the persecuted natives in the narrative, destined for demise before the oiled wheels (or machines) of progress.[10]

Fundraisers included a three-night 'melodrama' entitled 'Fair Barbara's Fatal Fault' or 'Oil's Well That Ends Well', that featured the villainous Derrick the Earl of Union. It raised $3,500. 'The Happy Time' musical benefit in April 1970 included a range of beach events and the cutting of a gasoline credit card, a small symbolic act of defiance. A GOO song sheet included ditties such as 'Dick, Please Check our Slick', aimed squarely at President Nixon; others focused on the black stuff and environmental squalor. Employing visual signifiers of protest, members identified with one another by sporting black 'mourning' bands and bumper stickers on their cars, with the rallying cry, 'Let Your Vote Be Heard. Sock it to 'em, GOO!' GOO also sought to spread a counter-narrative to the oil industry, popularising 'environmental education' in the guise of an information centre and leaflet drops. The protest group highlighted the negatives of the American oil industry for its audience: air pollution, industrial development, the loss of tourism and recreational economies, and the despoliation of nature.[11]

The message of GOO occupied a transitional zone, a liminal

space, between traditional conservation ideas and neophyte environmental thought. The original GOO mandate, 'dedicated to the promotion of the conservation of natural resources and the prevention of the pollution of natural resources', indicated this dual aspect to the movement. Preservation of nature seemed possible only by fighting head-on environmental dangers. The 'horror of polluting accidents' forced citizens to engage with the threat posed by modern contaminants to their environment. The oil spill serviced new environmental awareness.[12]

GOO featured an intellectual radicalism that often placed it outside the comfort zone of organisations such as the Sierra Club. Members criticised commercial development and some distrusted corporate America. GOO situated oil schemes as fundamentally 'incompatible with the environmental values we cherish', pitting industrial development as on a collision course with wilderness. The *Los Angeles Times* saw a 'battle shaping up' between old dollars and new greens. Nascent green thinking interfaced with civil rights discourse, with GOO keen to list complaints and demands in legalistic and politicised frameworks. Professor Norman Sanders explained, 'What we need is an environmental rights movement along with a civil rights movement'. GOO declared 'All men have the right to an environment capable of sustaining life and promoting happiness', referencing the Declaration of Independence. One resident explained that, 'every citizen has the right to a clean environment'. Their definition of rights incorporated a sense of environmental and community welfare. With its nod to local action and defence of home, GOO prefigured the sentiments of urban (social justice) environmentalism.[13]

GOO member and university professor Roderick Nash volunteered a set of eco-ethics, including an '11th commandment', to the group. The *New York Times* covered the story with the headline, 'Santa Barbarans cite an 11th Commandment: "Thou Shalt Not Abuse the Earth"'. The eleventh commandment imbued the movement with a sense of religiosity and self-righteousness. Nash went further than extant civil rights discourse, challenging conventional takes on social contracts. For Nash, rights could be broadened to include non-human agents of the natural world. His call to 'extend ethics' to a new ecological frontier smacked of radicalism and strove for a fresh 'ecological consciousness' of inclusivity. 'We will continue to fight for our environmental rights until they are recognized' declared the protesters in November 1970, while calling on the newly created Environmental Protection Agency (EPA) to investigate pollution. GOO called for a 'new green politics' in America. Local (and Sierra clubber) Fred Eissler wrote to the *New York Times*, 'We need a new politics – the politics of ecology – based on reverence for the land and reverence for man'.[14]

The 'Santa Barbara Declaration of Environmental Rights' forwarded a list of charges against humanity: litter, soil depletion, air and ocean pollution, poisons, extinctions, overpopulation and wilderness. It tapped all kinds of environmental concerns, both old, in the form of wilderness loss, and new, in the guise of chemical poisons. GOO thus recognised a range of environmental concerns beyond oil. Blame fell on the individual as much as on the industry. One member decried 'man's senseless exploitation and destruction of his own living environment!' while another, Theodore Schoenman, quoted Aldous Huxley, reflecting, 'We behave as though we were not members of the earth's ecological community'. Reminiscent of Rachel Carson's writings, GOO raised the issue of being 'out of equilibrium', that, 'Today we are living in such accelerated times that each and every hour compounds our struggle to keep in balance with nature'. Significantly, a 'doomsday' aspect featured in the declaration when it noted how 'centuries of careless neglect of the environment have brought mankind to a final crossroads'. For GOO, a 'do or die' situation seemed on the horizon, when 'The quality of our lives is eroded and our very existence threatened by our abuse of the natural world'.[15]

Much of the appeal of GOO lay in its timing, its focus on oil, its linkage to broader environmental topics, and its sense of place. The group resembled a 'community' project, with people and protest synonymous with place. As one local stated, 'Go get 'em Santa Barbara', and thanked protesters for their 'truly heroic efforts' at the time of the spill. GOO captured the anger and upset provoked by the spill, and served as an emotional outlet for a 'hit' community. 'If emotionalism is what it takes to make people stop and take notice of the rapid destruction of their environment, then I say to each and every man, woman and child: BECOME INTENSELY EMOTIONAL!' one resident railed. Politically, GOO garnered a wide spectrum of support. The Students for a Democratic Society (SDS) saw the spill as 'part of the general picture of tragedy which has befallen our American society' and supported the protest. Equally, *Newsweek* reported on how conservatives on the California coast had been 'radicalized' by events. In a 1970 survey of beach use, 173 out of 259 interviewed felt affected by the spill, with nearly half in opposition to oil platforms. Another survey noted 72 per cent of residents wanted oil out. A reporter for the *Washington Post* overheard one 'wealthy widow' in a local restaurant sharing how 'We ought to go out and burn the goddam things down', in reference to the oil rigs off the coast. As another activist highlighted, it was a simple decision, to 'stand up and be counted'. As a white, affluent, middle-class community, however, Santa Barbara lent GOO a decidedly limited patina in terms of race and ethnicity.[16]

The dystopic vision

Crucial to the success of GOO was the forging of a powerful negative vision, an eco-dystopia (or imaginative doomsday landscape) that frightened people and compelled them to act. Locals, press and activists all exploited the excesses of the oil spill to meld a forceful image of disaster. Environmental protest depends on a culture of fear and catastrophism in society to survive. With its dying birds on the blackened shoreline, the oil spill provided the material fabric for one potent doomsday vision of the 1960s. Adding to the dystopias of pesticide poisoning, nuclear apocalypse and the population bomb, the oil spill emerged on the shoreline of America as another new and powerful dystopia of the period: a realm of wreckage, decay and black days ahead. It provided a fresh fear factor, another eco-omen on the horizon.

This vision of dystopia drew on significant real-world inspiration and imagery. A starting point in any dystopia is the construction of Utopia: the fall of civilisation all the more powerful when presented with the peak. Press and protesters alike constructed Santa Barbara into an idyllic realm, a coastal paradise. One newspaper described it as 'one of the nation's most picturesque coastal areas'. Like conservationists enthusing over Yosemite and Yellowstone, protesters celebrated pre-oil Santa Barbara as invaluable and utopian. The coastal city symbolised greater California: golden, beautiful and rich. The oil spill jeopardised this stately image. Noticeable were the extremes at play: from Utopia to dystopia; from a pristine environment to a wrecked one; from paradise to disaster zone. 'We cannot maintain our environment and have oil development', declared one protester, reinforcing the bipolar imagery. The *New York Times* reported how Santa Barbarans 'see their environment and ecology threatened with destruction'. The oil spill allied with fears of fundamental transformation. Santa Barbara emerged as 'paradise lost'.[17]

This paradise was lost to one thing, the black stuff. The role of oil proved crucial to the presentation of ecological doomsday, of the metamorphic narrative. As the *San Francisco Chronicle* reported, 'The giant oil slick yesterday transformed the beaches of Santa Barbara into a nightmarish goo'. Beauty disappeared, engulfed by pollution. In the popular imagination, residents witnessed 'the loveliest residential and resort city', and the 'queen of all cities on the California coast' drowned in a 'nightmare' whereby 'greed for the almighty dollar has destroyed your beauty, has embedded you in a black oil slick'. This loss of the 'queen of all cities' resembled for some 'a death in the family', and smacked of 'ugly, senseless devastation'. Jinnie Huglin, a resident of Miramar Beach, captured the moment of shift: 'I find it hard to remember that a short time ago I was looking at clean sand – alive with birds, people and swimmers. Now – I see a wasteland.' Huglin lamented, 'I am now looking at black straw, black rocks, black driftwood, black waves, and black sand.'[18]

The trials of the oil spill encouraged Californians to envisage a town devoid of nature and beauty, an oily *Silent Spring*, and a place of encroaching blackness. Failing to see past the temporariness of the moment, reporters froze the dark image and published it. Two photographs featured on the front page of the *Santa Barbara News-Press*: the Union rig in the ocean and a ruined beach. The caption read: 'The oil disaster of 1969 – And what it did to Santa Barbara'. The article noted how 'Now, what has been feared for some time has happened! . . . ruined Southern California harbors and beaches . . . destroyed the area.' The greatest terror had been realised. One letter-writer nicknamed Santa Barbara 'Oiltown USA', and amended the words of a line from *America the Beautiful* to 'From sea to slimy [shining] sea'. More outlandish presentations of oil at Santa Barbara referenced the literature of medieval hell, 'The grimy fleet of duck boats, collecting gooey globs of straw within the breakwater, had reminded us of Dante's Inferno, the part where Charon ferries folks across the Styx'. Offering a more comedic turn on events, the *Santa Barbara News-Press* advertised the city as a reformed tourist haven. In a piece entitled 'Come One, Come All to Slick Town!', the newspaper predicted a range of 'slick and sophisticated fun spots' that included 'NEW tar pits', 'slick cruises to World Famous Platform A', and fishing for 'Petrole Sole'.[19]

In the process, Californians granted oil an agency and character of its own. To some extent, people anthropomorphised the black stuff to make sense of it, imagining 'oily fingers' of spoil menacing their coastline. The *San Francisco Chronicle* claimed that oil 'infested' the coast, as if Union had triggered an alien or pest invasion. No longer a simple chemical product, petroleum became intrinsically bad and dirty. Newspapers described an 'oil plague' unfolding while one resident wrote, 'Black gold is Black Death to Santa Barbara'. GOO constructed a sand wall on the beaches against the oil, to keep out what it labelled 'GOM' (the great oil monster). Talk of the 'stench of crude oil boiling and curdling', off the coast, of 'endless little slicks that sweat out from the metal monsters' gave oil a dark and malevolent character.[20]

In the popular retina, the oil fiend was inspired by classic horror imagery. On his presidential visit, Nixon faced placards imploring him to 'Ban the Blob', protesters equating the spill with the A-bomb (as in 'ban the bomb') while referencing *The Blob* monster of 1950s film fame. In the 1958 movie, Steve McQueen faced an ever-growing alien creature terrorising small-town America. The movie tag line described the Blob as 'Indescribable. Indestructible. Nothing Can Stop It.' Santa Barbara faced a similar threat. With 'almost heroic' efforts, citizens rallied with their straw and booms to limit the course of the enemy. As the *Sacramento Bee* reported, 'Men tried many things to stop the science-fiction like movement of the blob'. Sporting the ominous headline, 'Deadly blanket of darkness', the *Los Angeles Times* superimposed on the oil spill a malevolent nature, detailing how 'half an inch deep and an untold hundreds of miles in length and breadth, a black blanket of crude oil was still riding the long Pacific wells, spreading death and destruction'. The 'out of control' monster seemed intent on not stopping. Newspapers as late as June detailed how 8,000 gallons (30,000 litres) 'still ooze to the surface, every day, blackening the once-beautiful beaches, killing birds and marine life, fouling fishing and pleasure boats, and covering the water with an odorous slime'.[21]

There were also allusions to nuclear disaster in the dystopic lens. Covering the decline in the local fishing economy, the *Santa Barbara News-Press* described how a 'malignant oily monster', threatened the livelihood of the crew on the trawler *Atomic Girl*. The article conjured the image of an oil and nuclear disaster combined off the coastline, the *Atomic Girl* the oily mimic of the

Lucky Dragon Japanese trawler caught in the radioactive haze of Castle Bravo in 1954. Nuclear parallels ewmerged when considering the spread of the spill, the lake of oil travelling like a nuclear cloud, taken by nature and impossible to control by humans. Akin to a reactor breach or a nuclear test, the spill appeared without boundaries, unlimited, and uncontained. One reporter declared the oil project as outrageous as 'H bomb tests in the Grand Canyon'. Meanwhile, residents stemmed the tide as if fighting a radioactive monster from the movie *Them!* [22]

Radiation leaks and oil spills connected with a darker side of scientific discovery, a dystopia of technology run amok. The unravelling nature of nuclear- and oil-disaster scenarios underscored the fact that the machine could play havoc, that new inventions were rarely risk free, and that humanity had entered a new stage of technological doomsday. Humans had made the monster, created the Frankenstein. Bill Botwright, who reported on the *Torrey Canyon* spill, saw the Santa Barbara disaster as the plain result of 'an idiocy of man . . . idiocy, myopia and venality'. Reporters at Santa Barbara described the 'sticky problem', of fault lines, fissures and oil flow beyond the capacity of industry knowledge, of officials who 'still had no idea how to clean up the mess' days into the accident. The trial-and-error nature of clean-up techniques underlined the perilous situation. The use of straw and logs highlighted the rudimentary fashion of clean-up technology, the historical focus being on advancing drilling equipment and tapping oil, not limiting its flow. The *LA Times* claimed 'Southern California's oil disaster may become a classic example of modern man's failure to provide the tools to resolve the crises which he himself creates'. Technology failed to provide any solutions to the unfolding environmental quagmire. Before Congress, Philip S. Berry from the Sierra Club moaned, 'Present technology can not provide adequate guarantees against further spills, and techniques have not been developed to deal with spills when they do occur'. Senator Alan Cranston vowed, 'We must make certain that this situation never happens again'.[23]

The technological dystopia also tied with the impact of capitalism. The design of oil technology rested on the accumulation of profit. For Schoenman, the spill was an 'inevitable end product of a gargantuan and pitiless technology serving greedy economic expansion'. Rigs resembled evil transformers, giant machines at the behest of capitalist madmen. Casting the technology as immature and experimental, Huglin described how 'the oil platforms look like lighted, winking mechanical toys dropped at random in the sea by some giant child'. Like 'advancing fleet of battleships', the rigs threatened the California coast: Huglin's one hope was that a blanket of fog would smother them, and that the evil technology would disappear into the mist.[24]

Nature played two distinct roles in the disaster imagery: menace and victim. The casting of nature as menace added to the dramaturgy of the spill. Reporters placed Santa Barbara alongside natural disasters emanating from volcanoes, earthquakes and typhoons. One paper related how the disaster was 'like looking into the pit of an active volcano', while another claimed the fate of the coastline depended on 'the luck of the wind'. The reputation of California as 'earthquake country' cemented the doomsday image. Identified as the worst tragedy to hit Santa Barbara since the 1925 earthquake, the spill was identified with the region's seismic past. One reporter told: 'In 1925, nature readjusted small parts of her land mass in this area, in 1969 greed, exploitation and determination regardless of consequence has brought about a far more dreadful destruction.' The seismic potential of the Golden State spread fear 'of similar disastrous oil accidents' on the horizon. A GOO petition to Nixon foretold how 'one sizeable earthquake could mean possible disaster for the entire channel area'. With fault lines everywhere, GOO members confided, 'we shudder over the possible effects of a predicted, large earthquake on pipelines and platforms'.[25]

Casting nature as victim, meanwhile, fitted the image of Santa Barbara as paradise lost. The simple narrative followed: nature made the coastal paradise and then the oil industry destroyed it. Talk of the 'willful ruination of all the beauty which nature has lavished on Santa Barbara', of 'priceless and irreplaceable' natural beauty being cursed, underscored the 'paradise lost' analogy. Santa Barbara represented prime coastal wilderness defiled by the careless industrialisation of humanity: the spill a potent symbol of California Wilderness desecrated. Locals wrote of having 'experienced the beauties of the wilderness in their back country', a realm now threatened by oil. The growing spill tarnished the last true remaining wilderness of the ocean. In her 'Diary of Disaster', Huglin wrote: 'Man has always had one unshakeable belief: The earth might be wracked and torn and burned, but the vast and rolling sea was beyond our touch. It would always remain pure and undefiled.' Rachel Carson similarly shed light on threatened marine wilderness in her books. UCSB student James J. Joslyn updated Henry David Thoreau's canon, by noting 'in wilderness is the preservation of the world, but oil adds only to its corruption'. Californians recognised the fast disappearing wild coastline and the rise of new landscapes, shaped by technology and industry. 'Beaches will have become industrial parks or nuclear power plant sites,' feared one resident. The replacement of wilderness with sterile and concrete industrial landscapes seemed inevitable.[26]

Terminology also asserted the 'nature as victim' narrative. Santa Barbarans employed a range of emotive terms to dramatise impact. The coastline experienced 'the ravages of oil pollution', the 'flagrant violation of their environment', and 'the rape of the seashore' – such phrases played with notions of innocence lost and the conception of nature as female/victim. Popular in newspaper vocabulary, 'menace' implied a hostile nature to oil. The *Los Angeles Times* described 'a seething black scum'. The content of reports, meanwhile, focused on the imminent loss of rare and valuable ecology. Several articles highlighted fears for marine wildlife survival in the Channel Islands. Ranger reports of Anacapa surrounded by oil and 'trapped' sea lions and seals being 'forced into the oily ocean to eat and probably would ingest oil and die', captured attention, the 'dismal fate' broadcast in mainstream media. The *San Francisco Chronicle* reported on San Miguel Island and the 'growing ecological crisis' there. Images of oily pups and rumours of poisoned whales indicated the physical and biological scale of damage. Sierra Club steward Fred Eissler called the spill a 'major wildlife disaster'. The *San Francisco Chronicle* ran the controversial headline 'Island of Death'.

David Snell's article for *Life* magazine, 'Iridescent Gift of Death', proved particularly emotive. With black-and-white photographs of dying sea lions and blackened beaches, Snell foretold the end of nature off the California coast. A boat trip to San Miguel highlighted the widespread scale of damage, how normally 'a vessel such as ours would leave a fiery wake of phosphorescence in waters alive with microscopic marine life. But now the churning of our passage was dark in a sea gone dead.' At San Miguel, he discovered a wilderness haven blackened by oil, 'things not in balance' owing to a 'harsh and ugly a disruption as ever . . . inflicted upon a harmonious environment, the oil tide had altered many of nature's subtle mechanisms'. Snell was shocked by the dying sea lions on the island, detailing how 'we came upon oil-drenched pups that cried weakly and thrashed about like scalded rats, their eyelids gammed shut, umbilicals stained and caked'. Snell witnessed an environmental dystopia, having 'passed the calcified reminders of the primordial world that once was. Behind us was the black vision of the dead world which may come.'[27]

An oily dying bird inhabited the foreground of this environmental dystopia, giving it an iconography and a star. A total of 3,686 birds died because of the spill. Captain Cliff Matthews from California Fish and Game Department declared the Santa Barbara oil spill 'the biggest disaster ever to hit California's bird life'. State and national media published scores of photographs of oily and decrepit grebes, guillemots and gulls, multiple images imparting the scale of the dead that littered the coastline. The *New Orleans Times* told how 'Dead sea birds were washed up on blackened beaches Saturday as a vast oil slick continued to float unchecked in the Pacific Ocean'. Like miners' birds in cages, dead seabirds indicated the scale of the disaster. With a picture of an oily cormorant accompanying the story, the *LA Times* reported, 'sea birds died by the hundreds, mired in half-inch deep oil', that 'dead and injured birds cried weakly in the oil'. The image of oil-drenched birds dominated cartoons of the spill. A cartoon for the *San Francisco Chronicle* showed an oil worker with a rope around a very frightened, oil-soaked bird, 'The Ancient Mariner's albatross: Santa Barbara 1969'. Looking down on a blackened seabird, a Santa Barbara couple commented, 'Well . . . it looks as if we've just about pushed our environment to its limit'. The bird served as messenger, or symbol, of humanity inviting doomsday. A Hugo cartoon for the *Sacramento Bee* related the avian perspective: two oil-slicked birds on the beach reflecting on the clean-up operation before them, 'They're always sorry for their oil spills . . . and they're always sorry for the mess they've made – from a shorebirds viewpoint I'd say oil producers are just about . . . one of the sorriest industries I know!' Another sketch for the *Bee* showed rising anger between man and nature. In a dispute between sailor and gull, the sailor swiped, 'Don't Litter my Ship with Seaweed!' while the seagull dived, 'Don't Spill any more oil in my ocean!!!'[28]

In text and image, oil equated with the demise of avian life. The press employed the seabird as a visual motif of the oil disaster, fashioned it into the dominant symbol of death and decay. It simplified the broader picture of a coast threatened by environmental danger. The bird became the cultural signifier of environmental disaster, a totem animal, something to feel sorry for. In popular coverage, nature/bird translated as 'victim' throughout. Using transactional analysis and situating 'nature' as 'victim' within the Karpman drama triangle, the public assumed the role of 'rescuer'. The oil industry took on the third role, as persecutor. The press lavished attention on the fate of coastal birdlife, to the extent that some people complained that coverage valued avian species over oil workers. Scholars Carol and John Steinhart felt that 'The press capitalized on the slaughter of the birds. Wherever photographs of the oil spill appeared, there was a generous supply of views of oil-drenched birds, pitiful creatures already doomed but still struggling to free themselves from their black fate.' For the Steinharts: 'It was the plight of the birds that stirred the most feeling'.[29]

Part of GOO protest iconography, the bird appeared on banners, placards, cartoons and leaflets, and served as a motif of broader environmental problems. GOO members highlighted the links between their spill and wider pollution. A local artist, Anne Goetzman, produced an 'oil protest painting' of birds lost in blackness, the caption reading, 'Our vision is diluted by it. Our sea is polluted by it. Our birds are killed by it. Oil.' Further afield, citizens read of the collapse in seabird numbers and interpreted avian death as an environmental omen. One New Yorker linked avian demise to a whole range of maladies, claiming, 'The public rage is not just an outcry against the deaths of birds, however much these deaths grieve many, but an outcry against a symbol of what unrestrained development has produced'. Bird death

> may remind us that we too are threatened by the deadly air and water that we have contaminated ... We are part of a delicate balance of life, and the deaths of the birds in Santa Barbara show us that the scales are rapidly turning against us.

Hunter conservationists feared that the loss of seabirds was symptomatic of broader neglect of wildlife in the United States. The *American Rifleman* magazine labelled the oil spill a 'silent, creeping form of wildlife victimization', and encouraged readers to protest: 'The sportsmen of America can ill afford to remain silent while Black Death plagues our already beleaguered waterfoul' [*sic*]. Others connected the fate of oil-afflicted breeds with past species extinctions. 'The western grebe will have gone the way of the passenger pigeon and the California grizzly', one feared. Certainly stories such as

'Sea birds by the hundreds were reported dead or dying' facilitated images of mass death. For some, this led them to imagine landscapes of extinction, pictures of eco-holocaust and animal apocalypse. A headline in *The Nation* simply read 'Mankind's Fouled Nest'.[30]

The Santa Barbara oil spill became America's first modern 'environmental disaster'. Reporters, protesters and public alike treated the spill as an 'ecological' accident. The disaster interfaced with a range of environmental anxieties from nuclear war to pesticides. The spill provided an outlet for concerns and worries that had been brewing for decades over the environment. The psychological linkage between the oil spill and broader environmental issues proved significant. A pattern of neural pathways bonded around eco-decay: images of dead birds and ruined beaches collided with 'small planet' and big pollution ideas. Santa Barbara provided 'proof' that something was wrong. *Silent Spring* may have galvanised the public but it failed to offer a physical example of 'A Fable for Tomorrow'. Santa Barbara was the closest thing to the fable town realised.

GOO activists consciously situated the spill within this broader environmental dynamic. Making connections between oil and environment serviced the growth of GOO, gave its canvas national appeal, while also raising national environmental consciousness. GOO claimed to represent 'the American people'. A petition sent to Nixon argued 'Pollution and adverse exploitation of the natural environment are presently of extreme concern to people everywhere'. For UCSB student Joslyn, fighting Union Oil was part of a national battle. Joslyn explained how, 'Increasing population and progressive technology are helping to pull the bricks from man's foundation of survival – nature. This present unnatural oil leakage adds to the total "efforts" of man to destroy nature and thus himself.' Tying oil with the fate of the environment, protester terminology cast the spill as part of 'a chain'. Schoemann interpreted the spill as 'a further link in the chain of ecological suicide'. Activist Bill Denneen of Nipomo connected the oil disaster to all kinds of maladies, from suburbanisation and 'auto-geddon' to the urgent need for recycling. Denneen fretted how, 'Each year our environment becomes a little less fit for human habitation. Our insecticides and radioactivity are being concentrated in the food chain.' Again, the individual spectres of nuclear, pesticide and oil disasters metamorphosed into one huge eco-catastrophe.[31]

The press facilitated this process. 'Major Polluter of the Planet', read one newspaper headline, and placed the Santa Barbara spill alongside DDT and other concerns, as a 'piece of a larger, nation-wide picture which deserves broad public attention and action'. Cartoons carried an eco-pitch. In one, a Union Oil spokesperson explained to an unfriendly crowd (with placards 'Save the Birds' and 'Get Oil Out'), 'But look at the brighter side: None of that oil will be used to make smog.' An *LA Times* cartoon series depicted a drilling rig converted into a marine-based national park (named after Secretary of the Interior Walter J. Hickel). A banner on the side of the structure read: 'See the Nature Oil Seepage in Santa Barbara Channel (This observation platform provided by your friendly oil company).' In the next sketch, one whale-watching child from the rig asked his father, 'Dad, why are the gray whales all black this year ...?' The spurting of water from the whale's back resembled an oil fountain. The spill also coincided with the Apollo 11 moon landing and images of the Earth from space. The press reported 'nationwide concern' over environmental issues thanks to Santa Barbara and 'mankind's first dramatic view of planet earth from outer space'. One cartoon of the moon landing referenced the spill itself, 'call Mission control! ... tell 'em there seems to be an oil slick on our landing site'.[32]

The Santa Barbara spill emerged as a harbinger of a much greater dystopia ahead. Oil was part of a trajectory, 'humanity's rush to oblivion'. One Hollywood resident labelled the incident the 'most incredible of all shocking things – the unconscious efforts of man to organize his own mass self-destruction'. The *LA Times* declared the spill 'an ugly showcase of man's capacity to befoul his environment'. A cartoon showed a dustbin (marked 'pollution') leaching oil all over the globe.

Life magazine announced 'Suddenly we are all conservationists', that

> before long air pollution will screen out the sun and make big cities uninhabitable, that the fragile biosphere we all live in is becoming poisonous and may cease to support life, plagues threaten, the polar icecaps may melt and inundate us, or – take your pick – a new ice age may come.

Warnings to the Nixon administration of 'America uninhabitable within 10 to 15 years' bred a crisis mentality. The spread of the spill was widely taken as the beginning of an unstoppable flood of fresh environmental disasters, the true start of inviting doomsday.[33]

The villains

Protesters cast two villains responsible for the mess: the oil industry and the United States government. Many Santa Barbarans hardly welcomed the oil industry before 1969. The day of the leasing was cast as a 'black day in Santa Barbara County's history', when suspicious men 'many with Texas and Oklahoma drawls', came into town. One resident claimed that, 'For the majority of the people of Santa Barbara, the oil rigs in the Channel are the same as for the people of Czechoslovakia with Russian tanks in their country'. Another wrote simply, 'It's like an invasion from a foreign country.'[34]

After the spill, locals protested against the presence of oil companies with renewed zeal. *Newsweek* related, 'melodramatic David-&-Goliath story lines, with a doughty band of citizens pitted against giant oil companies'. GOO publications challenged the scientific authority of the industry. Activists presented the public with a choice over who or what to believe, ending the dynamic of singular authority, a step very much in line with the green agenda of the period (akin to Carson's *Silent Spring* and its challenging of the chemical industry). GOO called into question the oil industry's dystopic vision of oil running dry without fresh drilling, by suggesting there was no disaster on the horizon in terms of oil scarcity. Fears of oil running out within fifteen years were dubbed bogus. Oil supplies could last well into the twenty-first century.[35]

Criticism of the oil industry revealed a broader scepticism of corporate America. The *Santa Barbara News-Press* noted that, 'there are many people in this nation concerned about the Almighty Dollar's exploitation of God-given countrysides and seaways'. The nascent environmental movement sported anti-capitalist rhetoric. The handbook for Earth Day included an article on corporate 'green-washing'. Union Oil's 1968 announcement that 'All the platforms will be painted a soft blue to blend with sky and ocean' smacked of such propaganda.[36]

Many Santa Barbarans blamed the federal government for the spill and critiqued the ties between the Nixon administration and the oil industry. Presaging the Watergate scandal, protesters spread a story of conspiracy, and advertised 'The SBC oil situation' as a 'national scandal'. County supervisor George H. Clyde felt government officials made a 'grievous mistake' in granting the 1968 leases, angry that 'everyone in the Interior Department was hell-bent to get oil leasing in this area'. The perception of collusion between big oil and big government riled many. 'We Santa Barbarans bitterly realize that we are not living here in a democracy but under a dictatorship of oil interests,' read one letter. 'Beautiful Santa Barbara is being raped by political intrigue with federal politicians and slick oil men,' read another missive, 'slick tongued politicians' threatening the 'innocent' coastland.[37]

Congressional hearings revealed a deep-seated sense of betrayal. Theodore Schoenman claimed the spill was a perfect experiment in 'How to radicalize a conservative community within one short year'. GOO chairman Alvin Weingand provided evidence of the 'Department of Interior working hand-in-hand with the oil industry' in an 'unholy alliance'. Highlighting poor regulation, Senator Cranston claimed 'ample evidence of the pitifully weak Federal program for planning, developing and supervising oil production on the tidelands'. Emotions ranged from disappointment with the government's duty of care to stringent anti-federalism. Some residents denounced 'the federal government ravaging the coast'. A huge banner over the Route 101 overpass read 'Visit Santa Barbara's Dead Sea: A Project of your Federal Government'. For government officials, the oil spill caused much embarrassment. Former Secretary of the Interior Stuart Udall labelled the events in the Channel 'a conservation Bay of Pigs'.[38]

Environmental revolution

> We propose a revolution in conduct toward an environment which is rising in revolt against us.
> Santa Barbara Declaration of Environmental Rights, GOO

The 'conservation Bay of Pigs' was tied to popular revolution. Events in Santa Barbara connected with a national breakthrough in environmental consciousness. On 28 January 1970, Santa Barbara City College hosted a GOO discussion attended by the presidents of the Sierra Club, Wilderness Society, and National Audubon Society, alongside radical thinkers such as Paul Ehrlich and Roderick Nash. The city declared a day of concern for environmental rights. The event presaged the first Earth Day in April.

'GOO has become symbolic of environmental concern in the United States and is the best known citizen's group anywhere', claimed one commentator. It was an exaggeration, one perpetuated by GOO's own sense of mission and self-importance. But GOO spawned spin-offs such as KOO Portland, GOO Northwest, NO Pacific Palisades, and GOO Two in San Pedro. One resident of Whitefish, Montana wrote to the *Santa Barbara News-Press*, enthusing over 'similar crusades' across the country. Another noted how, thanks to Santa Barbarans, 'Everybody is talking about the environment'.[39]

A 1970 GOO press release read, 'This is not just local. The disaster which has afflicted our shores has become a nation-wide issue.' Before Congress, Mayor Gerald S. Firestone related that, 'even though we are a small city, we are the focal point I think for 200 million Americans today dealing with pollution'. City Attorney for Santa Barbara, A. Barry Cappello, labelled the event the starting point, or ground zero, for 'one of the major environmental and energy crises in the nation's history'. GOO director and UCSB professor Norman Sanders claimed, 'Santa Barbara calamity [as] a turning point in the campaign for a decent environment'. The local newspaper declared 'the nationwide environmentalist movement was born in this spilled oil'. For the *Christian Science Monitor*: 'It gave us the environmental revolution.'[40]

'Black Gold Blues'

The American oil industry responded to the Santa Barbara incident with a three-fold strategy: portraying the spill as a freak of nature (and thus reducing corporate accountability); playing down the scale of the damage; and offering an alternative disaster scenario more frightening than the spill itself.

President of Union Oil Fred Hartley proved vociferous in his defence of the industry. For Hartley, the spill signified an environmental anomaly. The trouble was simply 'Mother Nature is letting the oil come up'. According to the Union boss, 'The problem was that Mother Nature let the oil out of the drilling sands in a most unexpected fashion'. Before the Senate Sub-Committee on Air and Water Pollution meeting in February 1969, Hartley summarised his key points: that Union had entered into a partnership with the federal government and that the accident was beyond Union control or liability. More an act of God/nature than evidence of corporate malpractice, Hartley was 'sure everyone showed due diligence. Mother Nature is always showing us new things.' Put simply before the Senate, Hartley confided, 'We did not willfully pollute'.[41]

Secondly, the disaster was not a disaster. For Hartley, the term 'major event' seemed more appropriate. He explained, 'I think of disaster in terms of people being killed'. Hartley claimed that the real problem was 'pollution of certain elements of the media', with exaggerated reports of animal death and marine contamination. To the American oil industry, protesters and press were guilty of popularising a minor accident. *Oil and Gas* magazine detailed the 'numerous inaccurate, emotional, and oft times fabricated accounts of widespread damage ... Fanciful fiction'. Santa Barbara was nicknamed 'anti-oil country'. The oil industry promoted the idea of corporations as intrinsically 'rational' versus activists as resolutely 'emotional'. Harry Morrison, from Western Oil and Gas, cast it all as one big 'misunderstanding'.[42]

The spill also caused no harm to the environment. Within days of the spill, *Oil and Gas* reported that 'early damage to beaches and wildlife was amazingly light'. The magazine insisted nature, not oil, took responsibility for any mess. Another article described a bucolic 'forest' of petroleum platforms, whereby a blowout briefly combined with a flood, and that natural seepage overtook any artificial spilling of oil. Citing the work of Dr Wheeler North of Caltech, the industry pointed to the absence of fish or whale deaths as proof that the spill had caused 'no damage of any significance'. Funded by Union Oil, marine scientist Dale Straughan highlighted the problems in press coverage. Hartley bemoaned:

> I'm always impressed by the publicity the death of birds receives compared with the death of people. Relative to death that occurs from crime in our cities, the desecration of the offshore area of Santa Barbara – although its [*sic*] important and a problem we are fully devoted to – should be given a little perspective.[43]

Industry executives also highlighted the speedy recovery of the coastline. A journalist for Sun Oil underlined 'the final irony ... that where *Life* magazine had found a sea gone dead, I had found nature, and wildlife, in abundance'. Within two months, Union proclaimed, 'it looks as though the Santa Barbara area will be the top tourist recreation spot this year'. *Oil and Gas* enthused, 'Along

5 miles of clean, sunwashed beaches, swimmers splash in the gently rolling surf or loll on the sand . . . Even the Santa Barbarans acknowledge the beaches have never looked better.' Industry bosses welcomed the return of Santa Barbara and the resurrection of a Californian Utopia.[44]

Finally, the oil industry forwarded an alternative dystopia to that offered by environmentalists: that of a world run dry. The industry called on historic and enduring fears of oil running out, spouting energy declensionist tales. As early as 1926, the Federal Oil Conservation Board warned of a total collapse of oil resources within seven years. The imminent 1970s energy crisis proved a far more pressing issue than any environmental or aesthetic concern. One letter-writer to the local newspaper warned, 'Without these fuels the world would revert to worse than the Dark Ages', and relabelled GOO the 'GOOns'. The true dystopia was revealed as a barren energy landscape, a place without cars, electricity, or fires.[45]

Response and effects

The Santa Barbara oil spill led to significant discussion within the federal government. The President's Panel on Oil Spills produced two reports. The first admitted that the United States was ill prepared for any large spill. The report stated that the 'nation [was] not doing enough' about oil, and recommended more investment in technology, emergency planning, monitoring and regulations. The report pushed technological fixes hard, offering scientific ingenuity as the saviour:

> Although the development of a technology for handling spills in the open ocean is not the same order of magnitude as developing a technology to place a man on the moon or to build a nuclear energy system, the principle is the same.

The report also stated, 'In a sense, the oil spill problem joins the larger set of environmental problems facing society today'. The second report, however, claimed that the need for oil clearly outweighed the risks of further spills. Mineral resources on the continental shelf represented a 'new opportunity to add needed fuels, energy, metals and non-metals to our economy and monies to the gross national product'. No longer on a 'remote frontier'

Figure 5.2 The *Deepwater Horizon* oil spill: 'Fire boat response crews battle the blazing remnants of the off shore oil rig *Deepwater Horizon*, April 21, 2010' (US Coast Guard photograph).

but in coastal waters, this oil frontier simply required better safety protocols and high technology. The report also noted that 'The oil-soaked birds fluttering helplessly to almost certain death do not represent an ecological disaster'.[46]

The environmental impact of the oil spill gradually dispersed. The California Department of Fish and Game reported 'few deleterious effects . . . on the living marine resources'. Fishing, tourism and sightseeing gradually returned. *Time* magazine noted that 'initial fears' proved to be 'exaggerated', and the headline story of the Santa Barbara oil spill was nature's recovery. One resident declared 'The oil wells are a necessary evil', that Union Oil could be trusted, and that, ultimately, 'It's all part of progress'. Local fisherman Manuel Gorgita confirmed to the *Santa Barbara News-Press*, 'This is not the end of the world'.[47]

By the time of the OPEC crisis in 1973, little evidence remained of the oil disaster on the California coast. Despite protestations from GOO diehards (Weingand railed, 'The SBC has the same relation to the "energy crisis" as Yosemite National Park or Sequoia Redwoods have to a lumber shortage!'), the dominant message was one of keeping America's energy landscape going. While Santa Barbara avoided further oil development off its coast, America's addiction to oil continued.[48]

After the Santa Barbara oil spill, environmental thinkers warned of more disasters on the horizon. In the 1970s, the Steinharts predicted a spill in Alaska while GOO warned that the country would 'suffer at least one new oil spill of major proportions within the next 15 years'. The forecast proved to be roughly correct. In 1989, the *Exxon Valdez* spilled oil off the Alaska coast. In 2010, the *Deepwater Horizon* collapsed. As one commentator wrote about *Deepwater*, 'It was a disaster that many say was long in the making, was foreseeable, and almost inevitable.'[49]

The *Deepwater Horizon* spill

In February 2010, the semi-submersible drilling rig *Deepwater Horizon*, contracted by British Petroleum (BP), entered the Gulf of Mexico off the Louisiana coast. The United States government had recently granted leases in the Mississippi Canyon area, a deep ocean crevasse notoriously hard to drill. The Gulf represented a new frontier of oil and a new energy landscape. The *Deepwater Horizon* began drilling in Mississippi Canyon Block 252 of the Maconda prospect.

On 10 April 2010, a blowout occurred at the *Deepwater Horizon* rig, triggering a massive explosion of gas, oil, and concrete. Power died, and, unable to disconnect from the well, the rig began to sink. Flames lashed at the sky. Journalist Loren Steffy described the blazing rig as 'a forest fire of metal', akin to a 'giant funeral pyre' on the ocean. Eleven crew members died and seventeen were injured.[50]

By 27 April, the resultant oil slick had spread to 100 miles (160 km) across, and crept within 20 miles (32 km) of the American coastline. On 30 April, it arrived onshore. For three months (eighty-five days), the spill continued, until BP finally capped it. An estimated 4.9 million barrels (200 million gallons/780 million litres) of oil covered a stretch of ocean the size of the state of Delaware. The spill was the largest in American history. In an address to the nation, President Barack Obama declared the *Deepwater Horizon* spill 'a massive and potentially unprecedented environmental disaster'.[51]

The environmental disaster hit a region that had not yet fully recovered from another doomsday event, Hurricane Katrina. Louisiana faced another offshore onslaught, this time from oil, not weather. State Governor Bobby Jindal declared a state of emergency. Jindal despaired, 'This oil spill threatens not only our wetlands and our fisheries, but also our way of life.' Oil booms on the coast resembled the flawed levees at New Orleans, both designed to keep the hazard at bay. Memories of Hurricane Katrina resurfaced. Images of a ruined Gulf coast re-emerged. *Deepwater Horizon* served as HK2, and solidified the reputation of the Gulf region as a twenty-first-century doomsday landscape.[52]

The *Deepwater Horizon* spill closely followed the contours of the Santa Barbara spill. The same issues repeated: poor preparation for the emergency; chemical clean-up problems; talk of the unwelcome invader and corrupt corporations; use of the wildlife symbol; and accusations of government neglect. The BP spill was disaster on repeat. Nothing had been learnt from Santa Barbara, and the United States was still inviting doomsday.

Like the Santa Barbara Spill, *Deepwater Horizon* was promoted as an environmental story. Media images showed the polluted, blackened ocean and oiled, dying birds. Television documented struggles to save coastal waters. Louisiana residents feared the effect of the oil landscape on the fishing industry. The sinking poignantly occurred around the fortieth Earth Day. Reporters drew attention to the ecological worth of the Gulf region, classed by some the fifth most important in the world (in a census of marine life) and distinctive for its loop current, bird migration, and wetlands. Horror came in the number of victims: 6,147 birds, 613 sea turtles and 157 mammals (mostly dolphins). One commentator described the Louisiana wetlands as a 'giant ecological sponge' soaking up pollution.[53]

Within the environmental narrative, the oil industry again emerged as a disreputable force. Critics drew attention to the poor safety and environmental records of the industry, as well as the wider contribution of oil to global warming. Oil executives served as the poster-boys of an unsustainable Earth. One scholar called the twentieth century the period of 'oil destruction'. The *New York Times* referred to the 'deep and risky' practices of oil drilling in the twenty-first century. Drilling in the Gulf, the riskiest of locations, smacked of industry desperation for fresh oil and more profit. As at Santa Barbara, the industry played down environmental effects. Rebranding itself a green company in the 1990s and relabelled 'Beyond Petroleum', BP sported green credentials. BP executive Tony Hayward promised that 'The overall environmental impact of this will be very, very modest'. He pointed out that 'the Gulf of Mexico is a very big ocean'.[54]

As in the cases of Union Oil at Santa Barbara and Exxon at Alaska, BP became a scapegoat for the industry. Having cut corners on drilling, safety and emergency preparedness, the corporation appeared highly negligent. Critics pointed to a list of other BP-related accidents. In March 2005, an isomerisation unit used for making gasoline exploded at BP's Texas City refinery, near Houston, killing fifteen employees and injuring nearly two hundred. Fire services arrived, with Captain David Teverbaugh describing a 'nuclear holocaust' before him. The energy landscape of Texas City, 2,000 acres (790 hectares) of factories, pipes, trailers, barrels, and chemicals, suffered from a poor safety culture, outdated technology, and had been historically neglected by Amoco and BP. BP received a $50 million dollar fine under the Clean Air Act. BP also experienced safety issues at Grangemouth in Scotland and oilfields in Alaska. American corporations distanced themselves from the 'poor safety culture' of British Petroleum. Exxon asserted that it simply didn't drill that way. The *Washington Post*, however, reported a record 12,087 oil-related incidents in the Gulf caused by the wider industry.[55]

Government monitoring seemed little improved from the 1960s. Established in the Reagan anti-environmental culture of the 1980s, the Minerals Management Service (MMS) officially oversaw drilling in the Gulf. As awarder, regulator, and enforcer of drilling, the MMS suffered from a conflict of roles. Staffed by oil-industry workers, and maintaining a close relationship with corporations, the MMS offered poor supervision. The Atomic Energy Commission had been split up in 1973 because of such ethical issues. With increasingly sophisticated technology, a 'knowledge gap' emerged between industry and inspectors. BP documents to support the Gulf project exhibited significant flaws yet still passed. An approved BP Oil Spill Response Plan for the Gulf was later described as a 'fantasy document' in its approach to rescue, while a BP Environmental Impact Statement (EIS) was identified as a recycled corporate template, citing a dead wildlife expert as witness and wrongly listing animals such as the walrus in the Gulf.[56]

The spill also exposed the agenda of politicians trapped in an outdated energy landscape. Despite increasing dependence on foreign oil (from 36.1 per cent in 1974 to 66.2 per cent in 2009), American politicians blindly chased the old chimera of energy independence, a broken pipe dream of the 1970s, when the Trans-Alaska Pipeline (TAP) suggested a slim chance at fighting OPEC dominance. Renown for its 'oil friendly' politics, the Bush administration (2001–08) contributed to the oil bias. Author Naomi Klein related the enduring pro-oil fervour of Republican Party conventions after 2008, of 'Drill Baby Drill', and a 'party base was in such a frenzy for US-made fossil fuels, they would have bored under the convention floor if someone had brought a big enough drill'. The Democrats also appeared guilty. In March 2010, President Obama opened federal water off the eastern Gulf of Mississippi and new parts of Alaska to oil, appeasing the industry (in 'trade' for support for a 'cap and trade' bill in the Obama green energy project). Just three weeks prior to the spill, Obama reassured the nation, 'It turns out, by the way, that rigs today generally don't cause spills. They are technologically very advanced.'[57]

Environmental disaster also connected with technological failure. In terms of drill design, Gulf rigs proved considerably more advanced than the rigs at Santa Barbara. Unlike Drake's first rudimentary salt-well driller at Titusville, modern drilling involved tension cables, thrusters and computer-controlled positioning. Akin to high-technology frontier cities, rigs in the Gulf sported the latest computers and support systems. Built by Transocean in 2001, at a cost of half a billion dollars, *Deepwater Horizon* had a crew of 126, featured powerful thrusters, and advanced mobile technology. It followed in a line of sophisticated exploratory rigs, such as the *Thunder Horse* platform (nicknamed 'a Monster'). Transocean's slogan of 'We're never out of our depth' oozed confidence, with *Deepwater* lauded as a feat of high-tech engineering. In 2009, it drilled the deepest well in history, 35,000 feet (10,670 m) down at Tiber Field. The Gulf environment proved very different from offshore drilling at Santa Barbara, let alone the land derricks of *There Will be Blood*. A new frontier of technology and geography promised more of a very old resource. But ultimately, as at Union's Platform A, the drill technology failed.

With warning signs ignored, poor management culture, human error and risky technology in evidence, the *Deepwater Horizon* incident proved reminiscent of technological disasters at Three

Mile Island nuclear plant (1979) and onboard NASA's space shuttle *Challenger* (1986). An allegedly failsafe backup device, the blowout preventer, failed at *Deepwater Horizon*. At Three Mile Island, a blowout valve accidentally shut down. As one scholar commented, 'The Deepwater Horizon shares many of the characteristics of other technological disasters.' The industry expected high technology to deliver, even when put into risky situations, such as tackling hurricanes. Faith in the machine endured, even prospered, in such conditions. 'Exploration for oil in the Gulf of Mexico has become ruled by the engineers' conceit that the industry's technology was impeccable', noted Loren Steffy. Industrialists claimed: 'it's possible to build something to deal with just about any conditions.' Confidence translated into ignorance. On the eve of the accident, Halliburton cement well casing, dubbed 'a marvel of technology', was clearly not up to the task, while BP reduced its centralisers (used to keep the drill in place) from twenty-one to just six in number.[58]

In terms of clean-up, far less progress had been made on the technological front. While drilling boasted high technology, clean-up featured the lowest. Design focused on state-of-the-art drilling equipment, frontier technology, while clean-up was left to straw bales. Despite two major spills at Santa Barbara and Alaska, no real investment had been made in advancing clean-up technology. Containment strategies offered little innovation. The industry also seemed overly confident, with BP claiming to be able to clean up an 'Exxon Valdez' per day (10.5 million gallons/40 million litres) when it struggled with a 60,000 barrel (2.5 million gallon/9.5 million litre) spill, let alone 250,000. Attempts by BP to innovate with technology and shut off the spill using a 'dome' then a 'top hat' both failed.

While media, government and industry briefly focused all attention on the *Deepwater Horizon* spill, once the disaster passed, a broader atrophy of vigilance returned. Environmental disasters represented fleeting diversions (and oil spills cyclical repeats) when the quest for oil remained constant and fixed. The American nation still relied on huge amounts of oil to fuel auto-culture and energy use. Oil remained very much part of the modern American energy landscape.

Notes

1. 'On the map', *The Titusville Herald*, 3 May 2011.
2. *Oil & Gas Journal*, 12 August 1968, MSS 11, 23/208, SBHC University of California, Santa Barbara Library,(herein all codes refer to UCSB collection); 'Black Gold Blues'; *Forbes*, 15 February 1969, MSS 11, 23/2.
3. 'Huge Slick off California Spreads', *New Orleans Times*, 1 February 1969, MSS 11, 1/1; 'Spreading Oil Slick . . .', *Los Angeles Times*, 3 February 1969 MSS 11, 1/1; Teague: 'Rebuttal to KIST Editorial', 13 May 1969, MSS 10, 21.
4. 'Slick is spreading . . .', *San Francisco Chronicle*, 1 February 1969, MSS 11, 1/1; 'Desperate Fight . . .', *San Francisco Chronicle*, 5 February 1969, MSS 12, 1/1; 'Thou Shalt Not Abuse the Earth', *New York Times (NYT)*, 12 October 1969, MSS 10, 25.
5. 'Santa Barbara's Ordeal by Oil', *Sierra Club Bulletin*, March 1969, MSS 11, 23; Advertisement in *Wall Street Journal*, 4 March 1969, MSS 11, 1; Carol and John Steinhart, *Blowout* (Pacific Grove: Duxbury Press, 1972), p. 80; reprinted *Los Angeles Times* cartoon in *Santa Barbara News-Press (SBNP)*, 29 June 1969, MSS 11, 54.
6. A. Barry Cappello, Santa Barbara Oil, January 1974, MSS 7, 4/6.
7. 'Oil Filth', *San Francisco Chronicle*, 6 February 1969, MSS 11, 1/1; House Representatives Committee on Public Works: Subcommittee on Harbors, Rivers and Pollution and Subcommittee on Flood Control (Santa Barbara, 14 February 1969) report, 48, Government Information Center, UCSB.
8. GOO press release, 5 January 1970, MSS 10, 21; Letter from Thomas Quinn, March 1969, MSS 11, 23.
9. GOO letter to Senator Edmund Muskie, March 1969, MSS 8, 1/5; GOO Constitution Article 2, GOO pamphlet, MSS 10, 23.
10. *SBNP*, 14 July 1969, MSS 11, 54.
11. GOO song sheet, MSS 10, 23.
12. GOO pamphlet, MSS 10, 23; GOO petition (undated), MSS 10, 23.
13. *SBNP*, 10 February 1969, MSS 11, 4/1; *Los Angeles Times*, 2 February 1969, MSS 11, 1/1; *NYT*, 12 October 1969; GOO, 'The Santa Barbara Declaration of Environmental Rights', MSS 10, 25.
14. *NYT*, 12 October 1969; *SBNP*, 20 November 1970, MSS 10, 21.
15. 'Declaration'; James Bottoms, GOO, MSS 10, 21; Theodore Schoenman, 'Link in Chain of Ecological Suicide', *SBNP*, 11 February 1969.
16. *SBNP*, 6 February 1969, MSS 11, 4/1; Bottoms, 'SDS View Presented on Channel Outrage', *SBNP*, 13 February 1969; *Newsweek*, 12 August 1969 MSS 11, 23; Post reprint in *SBNP*, 6 July 1969, MSS 11, 54.
17. 'Exploitation vs Environment', *SBNP*, 7 March 1969; County Superior George H. Clyde, House of Representatives Committee; *NYT*, 12 October 1969.
18. *San Francisco Chronicle*, 6 February 1969; *SBNP*, 3 February 1969, MSS 11 4/1; *SBNP*, 11 February 1969 MSS 11, 4/1; Jinnie Huglin, 'Diary of a Disaster', *McCalls*, June 1970, MSS 11, 53/2.
19. *SBNP*, 7 March 1969 front; Mrs Tom Gilliland, 'A Problem with Anthem', *SBNP*, 15 February 1969; 'Requiem for Beautiful Beach', *SBNP*, 9 February 1969; 'Come One, Come all', *SBNP*, 12 March 1969.
20. 'Slick is spreading . . .', *San Francisco Chronicle*, 1 February 1969, MSS 11, 1/1; 'Oil filth', *San Francisco Chronicle*; *SBNP*, 1 June 1969, MSS 11, 54; *SBNP*, 11 February 1969, MSS 11, 4/1; Bill Botwright, 'Oil vs Public', *Boston Sunday Globe*, 9 February 1969, MSS 11, 1/1.
21. 'Santa Barbarans Feel Personal Loss', *Sacramento Bee*, 9 February 1969, MSS 11, 1/1; 'Deadly Blanket of Darkness', *Los Angeles Times*, 9 February 1969, MSS 11, 1/1; *SBNP*, 1 June 1969, MSS 11, 54.
22. 'Chain Reaction', *SBNP*, 7 February 1969; Botwright, 'Oil vs Public'.
23. Botwright, 'Oil vs Public'; 'Oil Leak Presents . . .' *San Francisco Chronicle*, 2 February

1969 MSS 11, 1/1; 'Oil Slick Grows', *Sacramento Bee*, 1 February 1969 MSS 11, 1/1; 'Disaster Shows Man's Failure . . .', *Los Angeles Times*, 7 February 1969, MSS 11, 1/1; *Santa Barbara Oil Pollution Hearings* (1970), pp. 144–6; Alan Cranston, CA (senate) to Harriette von Breton, 29 March 1969, MSS 8,1/1.

24. Schoenman; Huglin.
25. *SBNP*, 29 January 1969; *New Orleans Times*, 5 February 1969; *San Francisco Chronicle*, 6 February 1969; GOO Petition to Nixon, MSS 8, 1/5.
26. *SBNP* letter Gertrud Heymann (UK), 23 February 1969; *SBNP* letter Aaron Armstrong (Goleta, CA), 4 February 1969; *SBNP* 12 February 1970 editorial; Huglin (1970); James Josyln, 'Pulling Bricks from Foundation of Survival', *SBNP*, 13 February 1969; *SBNP* letter M. Rodriguez (SB), 5 February 1969.
27. GOO mailshot, 30 October 1972 MSS10, 22; *SBNP*, 7 May 1969, editorial; Letter Eleanor Wright (SB), *SBNP*, 6 February 1969; *NYT*, 4 February 1969; *Los Angeles Times*, 6 February 1969; *San Francisco Chronicle*, 5 February 1969; 'Island of Death', *San Francisco Chronicle*, 3 April 1969; David Snell, 'Irridescent Gift of Death', *Life*, 13 June 1969.
28. Eloise Dungan, 'Let's go to Work', California Living, *San Francisco Examiner & Chronicle*, 6 April 1969; *New Orleans Times*, 2 February 1969; *Los Angeles Times*, 6 February 1969; *San Francisco Chronicle*, 7 February 1969 MSS 11, 4; *Sacramento Bee*, 23 February 1971 and 15 September 1970, MSS 11.
29. Steinhart (1972), p. 77.
30. Anne Goetzman, shown at Austin Gallery, SB; letter, Edgar Roberts, *NYT*, 23 February 1969; Jerome Knap, 'The Black Blight', *The American Rifleman* 117/5 (May 1969); Rodriguez letter; *San Francisco Chronicle*, 6 February 1969; Edward Cowan, 'Mankind's Fouled Nest', *The Nation* (10 March 1969).
31. *SBNP*, 24 April 1970; Josyln; Schoenman; Letter Bill Denneen (Nipomo), *SBNP*, 17 March 1969.
32. *SBNP*, 10 January 1969; Union cartoon, 1 April 1969, MSS 11, 4; *Los Angeles Times*, 21 December 1969 and 19 October 1969, MSS 11, 4; *SBNP*, 3 October 1969.
33. Schoeman; Oblivion; letter, Patrick Mahony (LA), *SBNP*, 11 February 1969; *LA Times* reprinted in *SBNP*, 25 December 1969; *Life* magazine, 'The Rhetoric of Ecology'.
34. *SBNP*, 8 February 1970; Alexander Pagenstecher, SB, *Hearings before the Subcommittee On Minerals, Materials and Fuels of the Committee on Interim and Insular Affairs* (Wash: 1970), p. 72; *Newsweek*, 18 August 1969.
35. *Newsweek* 18 August 1969.
36. *SBNP* editorial, 7 May 1969; Union Oil Press Release, 10 September 1968, MSS 7,4/5.
37. GOO telegram to President Nixon, 9 January 1971, MSS 10, 21; Clyde in *San Francisco Chronicle*, 6 February 1969; letter, Nancy Mahorga (SB), *SBNP*, 3 February 1969; letter, William Holmer, *SBNP*, 11 February 1969.
38. *Santa Barbara Oil Pollution Hearings* (1970), pp. 79–83, 135–7; Cranston letter; *San Francisco Chronicle*, 3 February 1969; *Oil and Gas Journal*, 17 March 1969, MSS 11, 23.
39. Scuttlebutt, SB Yacht Club (1970) MSS 8, 21; letter, Jack Hoterman (MT), *SBNP*, 19 February 1970; letter, Eileen Root (VA), *SBNP*, 22 February 1970.
40. GOO press release, 24 February 1970, MSS 10, 21; Hearings (1970); Cappello (1974); *SBNP*, 7 May 1970; *SBNP*, 7 May 1981; Mark Stevens, 'Santa Barbara Environmentalism . . .', *Christian Science Monitor*, 1 February 1979.
41. Testimony of Fred Hartley, president, Union Oil, before Senate, 5 February 1969, MSS 7, 4/10; Fred Eissler, 'Santa Barbara's Ordeal by Oil', *Sierra Club Bulletin*, March 1969, MSS 11, 23.
42. *Sacramento Bee*, 13 February 1969; letter to employees, Hartley, May 1969, MSS 11, 36; *Oil and Gas Journal*, 67/34 (25 August 1969) and 68/12 (23 March 1970), MSS 11, 23; Harry Morrison, 'Effect of Channel Oil Spill', *Los Angeles Times*, 26 April 1969.
43. *Oil & Gas*, 10 Feb 1969, MSS 11, 23; Morrison; Hartley, in "Is there a Ladder of Desecration," *SBNP* 12 Feb 1969.

44. Robert Klaus, 'In the Case of Santa Barbara', *Our Sun* (1969), MSS 11, 24; Union Oil (Hartley) to all employees, 4 April 1969, MSS 11, 36; *Oil and Gas* (25 August 1969).
45. Letter, Walter Hinzen (SB), *Goleta Advisor*, 5 February 1969.
46. 'The Oil Spill Problem', First Report of the President's Panel on Oil Spills (Washington, DC, 1969); 'Offshore Mineral Resources', Second Report of the President's Panel on Oil Spill (Washington DC, 1969).
47. *SBNP*, 5 March 1970; 'The Environment Not So Deadly', *Time*, 13 June 1969; 'Local Man on the Street . . .', *SBNP*, 9 February 9 1969; Tom O'Brien, 'Requiem for Beautiful Beach', *SBNP*, 9 February 1969.
48. Alvin Weingand to *Los Angeles Times*, 16 May 1973, MSS10, 21; Robert Sollen, 'Oil Spill fueled environmental movement', *SBNP*, 28 January 1979, MSS 10, 26. On the ten-year anniversary of the spill, the local paper took pride in how the 'Oil spill fueled environmental movement', offering 28 January as the 'anniversary date' of modern environmentalism. The spill, *Torrey Canyon* and 'the widely read warnings by the late Rachel Carson, aroused a nationwide public indignation that became persistent'.
49. See Steinhart (1974); GOO Policy Statement 1984, MSS 10, 21; Stanley Reed and Alison Fitzgerald, *In Too Deep: BP and the Drilling Race that Took It Down* (Oxford: Wiley, 2011), p. 17.
50. Loren C. Steffy, *Drowning in Oil: BP & the Reckless Pursuit of Profit* (New York: McGraw-Hill, 2010), pp. 10, 18.
51. President Obama, White House Press Release, 2 May 2010.
52. Bobby Jindal, 1 May 2010, *BBC News*, 2 May 2010.
53. Steffy (2010), p. 207.
54. William Freudenberg and Robert Gramling, *Blowout in the Gulf: The BP Oil Spill Disaster and the Future of Energy in America* (Cambridge, MA: MIT Press, 2011), p. 4; *NYT*, 11 August 2002; 'BP Chief: Oil Spill Very Modest', *Sky News*, 18 May 2010; *Guardian*, 14 May 2010.
55. Steffy (2010), pp. 78, 214; Freudenberg (2011), p. 55.
56. Steffy (2010), p. 201.
57. Naomi Klein, 'Gulf Oil Spill: A Hole in the World', *Guardian*, 19 June 2010; Steffy (2010), p. 194.
58. Freudenberg (2011), pp. 15, ix; Steffy (2010), p. 205.

CHAPTER SIX

The Disaster City and Hurricane Katrina

On 16 December 1884, the gates opened to the World's Fair at New Orleans, Louisiana. Thousands of visitors made their way by railway, pavement and even by steamboat to the fair. The exhibition space stretched from St Charles Avenue to the Mississippi. The main building spanned 33 acres (13 ha), the largest roofed structure of its time. Testament to the Thomas Edison era, 5,000 lights shone inside, some ten times the number located in the city outside. A giant Mexican brass band entertained while spectators explored electric lifts, Venetian glass, a Japanese tea pagoda and a refrigerated display of 10,000 packs of butter.

Geographic confluences of imagination, high technology, and internationalism, World's Fairs excited spectators with dazzling displays of the future. The most famous incarnation, Chicago's World's Columbian Exposition of 1893, testified to the hopes of the American nation. Captured within Chicago's 633-acre (249-ha) Jackson Park was a veritable cornucopia of technological modernism. Chicago's Machinery Hall featured its own power plant, providing power for a legion of singing sewing machines and turning the world's first conveyor belt. The agricultural building included a supersized cheese from Canada and a liberty bell made from oranges. Fairs abounded with urban fantasies and promoted whitecity templates of design. Fair architects designed future American cities within their parks, envisaging places of abundance, cleanliness and high technology. Some 27 million visitors attended the 1893

Chicago Exposition, all enthralled at the future mapped out for them.

But the fairs also existed within real American cities sporting significant social and environmental problems. Dubbed the World Cotton Centennial, the 1884 Louisiana incarnation celebrated the first recorded export of the crop to England in 1784, and the formative rise of the South's Cotton Kingdom. Aside from the ornamental electricity, the fair mostly seemed trapped in agriculture and the region's slavery past. Exhibits on the contributions of 'the colored race' and 'women's work' offered some sense of pride but without challenging the historic social hierarchy of the region. Funded by $50,000 apiece, the two exhibits displayed the handiwork and inventiveness of women and blacks but also asserted a narrative of separation, the exhibit sign 'Colored Department' a sound linguistic clue of lingering prejudice. The New Orleans example hardly set the tone for a white-city future of high technology and progressive egalitarianism. Corruption was rife; attendance at the fair disappointed. In 1885, city officials recycled the same fair structures for the North, Central and South American Exposition. New Orleans seemed trapped in its colonial, racial and agricultural malaise.[1]

The white city was not just a mirage at New Orleans. Most other cities failed to deliver a decent quality of life at the turn of the century. The reality of American urban residence proved far less comfortable than projected. Real American cities proved far less salubrious than fair models and exhibits. Chicago was a case in point. The Chicago Fire of 1871 wiped out 4 square miles (10 sq. km) of the city, with a toll of 17,500 buildings and several hundred lives. The industrial, meat-packing metropolis that rose from the ashes became the source of much consternation. Socialist writer Upton Sinclair captured the sense of Chicago and, more broadly, the American City as resident evil. He cast the city as *The Jungle*, a tangle of corruption, disease and profit-making. With vitriol and wit, Sinclair described the horrors of basic processes, such as food manufacture, in the urban domain. The typical American sausage produced in Chicago factories was a 'moldy and white' European reject, full of rat dung and sawdust, and occasionally 'smoked' with borax to disguise the taste of 'dirt and rust and old nails and stale water'. According to Sinclair, 'there were things that went into the sausage in comparison with which a poisoned rat was a tidbit'. Sinclair's 'tubucular' sausage was a masterpiece of public health scare and socialist criticism. The turn-of-the-century American city was a decidedly unsavoury place: unnatural, dirty, and capable of spreading contagion.[2]

While Sinclair documented the sausage scandal, Jacob Riis exposed the horrors of tenement overcrowding in New York City. An exercise in early photojournalism, Riis documented the problems of mass tenement housing in the eastern metropolis, the 'hotbeds of the epidemics that carry death to rich and poor alike.' He depicted a world of tramps, beggars and poor values, that 'touch the family life with deadly moral contagion'. Riis highlighted issues of infant mortality, poor public (and landlord) education, and overpopulation. The book was richly illustrated with snapshots of squalor. Riis felt that tenement-based New York City was an abomination, a monster: 'We know now that there is no way out; that the "system" that was the evil offspring of public neglect and private greed has come to stay. A storm-centre forever of our civilization.'[3]

At the root of the urban crisis lurked environmental ignorance. The sterile, anti-nature project of the American city was found seriously wanting. Sociologist Lewis Mumford noted the failure of the city experiment whereby 'mechanical processes had supplanted organic processes'. Limits went ignored. Cities were unhealthy, overpopulated and stinking places. Rather than promising a great future for citizens, urban structures instead provided an unhealthy and despoiled environment. Ignorance of the natural environment produced catastrophic results. Building in the wrong places led to open invitations of doomsday. In San Francisco, the 1906 earthquake and fire wrecked the city. Author Jack London described the scenes of devastation for *Collier's Magazine*, highlighting the failure of the modern technological, industrial metropolis to cope with natural forces:

> All the cunning adjustments of a twentieth century city had been smashed by the earthquake. The streets were humped into ridges and depressions, and piled with the debris of fallen walls. The steel rails were twisted into perpendicular and horizontal angles. The telephone and telegraph systems were disrupted. And the great water-mains had burst. All the shrewd contrivances and safeguards of man had been thrown out of gear by thirty seconds' twitching of the earth-crust.

'Not in history has a modern imperial city been so completely destroyed', declared the author. In response to the earthquake and fire, London renamed San Francisco 'the doomed city'.[4]

The doomed city project

By the 1970s, the ongoing developmental challenges of the American city had combined with the advent of modern environmentalism and given rise to all kinds of doomsday fixations. Images of overcrowding, corruption, technological malfunction, all loomed over the cityscape. Urban health issues conjoined with burgeoning environmental fears over finite resources, population growth, smog and industrial pollution.

Environmental doomsday texts, such as Paul Ehrlich's *The Population Bomb* (1968), facilitated mass anxiety over the cityscape. Ehrlich made a simple argument of 'carrying capacity': that world population was vastly outstripping resources. In emotive terms, he declared: 'the battle to feed all humanity is over'. For impact, he put forward a range of sensational 'end scenarios', of fictive doomsdays ahead. Ehrlich provided a neo-Malthusian statement of population concern, of numbers outpacing supply. He added a fashionable environmental dynamic by linking the book with contemporary images of a fragile planet. English botanist Arthur Bunting called him the prophet of doom. In 1948, Fairfield Osborn had explored a similar threat in *Our Plundered Planet*. Osborn warned how 'we are moving towards the twilight of our civilization', that 'another century like the twentieth, civilization will be facing its final crisis'. Lewis Mumford warned of 'the slavery of large numbers', that cities offered a 'violent experience' as concentrations of bad power and people.[5]

According to Ehrlich's vision, American cities of the late twentieth century were about to be overrun with people. Ehrlich rendered the modern American city an imminent disaster zone and a potent symbol of an overburdened planet. Cities functioned as the launch sites for the coming population explosion. Mere decades earlier, the nuclear arms race and the Cold War had first put the American city on the frontline of danger. Urban centres represented targets on a military map, places of potential disaster, regions literally to wipe out. They also provided places of shelter: the perfect nuclear symbol found at

Abo Elementary School, Artesia, New Mexico, an underground bunker-cum-school. Fear of radiation and Soviet attack placed the city at the ground zero of American anxiety in the 1950s. Ehrlich updated the 'ticking' bomb motif, situating the life of the city as running out because of a new demographic force. Targets for Soviet nuclear missiles, American metropolises were now the 'ground zeros' of the population bomb as well.

Science fiction films of the period captured this narrative of urban doom. *Soylent Green* (1973) recounted the exploits of sweaty city cop Thorn (Charlton Heston) investigating the strange new food rations distributed on the future east coast. The opening credits of the film documented the fall of humankind and city alike by a series of photographic shots of waste sites, food riots and urban damnation. The cityscape of *Soylent Green*, a dirty and overcrowded New York, was situated just on the horizon, in 2022. In this future New York, people died in the streets of hunger, or volunteered to end it all in the comfort of hi-tech euthanasia clinics, where a fatal injection was accompanied by rare photographs of nature, wilderness and happiness – a lost world. City residents also unknowingly ate each other: Soylent Green fed the masses (just about) but was also, quite literally, made from them. Green was the title of the food, and also the message. *Soylent Green* was an Ehrlich-inspired 'end scenario', a vision of environmental doom at the ground zero city. Other films projected the problematic cityscape into the future. Based on a Philip K. Dick story, Ridley Scott's *Blade Runner* (1982) documented a grimy 2019 Los Angeles overpopulated with humans (and six too many androids requiring termination). *Logan's Run* (1976) depicted a domed survival city in the twenty-third century, with resources and population in balance courtesy of a policy of compulsory termination at the age of thirty (with the process disguised as a psychedelic-futurist carousel ritual of vaporisation). The city endured as the epicentre of dystopian visions.

Did real American homes of the late twentieth century support any of this? Looking at the 2000 census, Patrick Simmons detailed a significant rise in overcrowding in post-suburbia America, with the problem increasing since the 1980s. He charted 6.1 million overcrowded houses in 2000. But there were other environmental issues at work rather than sheer population figures. Pollution blanketed the great metropolises, especially those designed around the automobile. While the 'vertical cities' of New York and Chicago dominated the late nineteenth century, in the late twentieth century, 'horizontal' cities, such as Los Angeles and Las Vegas, set the trend. The upward spiral of population and capital captured in the American-engineered skyscraper had been replaced by outward sprawl, while the car replaced the train as transport.[6]

A (post)-modern city shaped by the freeway, Los Angeles lacked sidewalks, public transport, integrated community or a clean environment. Lewis Mumford described Los Angeles as a 'motor-ridden conurbation'. David Cronenberg's filmic version of J. G. Ballard's auto-erotic, auto-suicidal text *Crash* (1973) explored the cultural auto-geddon associated with Los Angeles. In the film *Falling Down* (1993), William Foster (Michael Douglas), overwhelmed by smog, heat and frustration, embarked on his own individual 'fall' into hell by abandoning his car in a traffic queue on the LA freeway. The film touched on the everyday horrors of the morning city commute, and the 'pollution brown' horizon on every driver's windscreen in LA. The *Terminator* films (1984–2009) similarly projected a city destined for destruction, the origin point for a forthcoming apocalyptic war of the machines. Los Angeles symbolised a city on the edge of doomsday at the end of the twentieth century, and a sprawling architectural statement of environmental ignorance. Its constant outward growth reflected traditional American values of abundance and unlimited progress. But, as a desert city patrolled by the automobile, it faced the double threat of both oil and water running dry.[7]

A short drive from Los Angeles, Las Vegas served as the ultimate city defying limits, an architectural statement of environmental denial. A sprawling metropolis in the 100 °F desert, Vegas blatantly flouted the laws of nature. Fountains at Bellagio's Casino artificially mirrored the spurt of Old Faithful at Yellowstone National Park, streams of water cascading into the sky every thirty minutes in the daytime (as opposed to sixty-five minutes at Yellowstone). Crowds gathered to look at both scenes but, at Vegas, the fountains amounted to proud statements of environmental defiance. The theme park Vegas Strip was a wonder of fantasy and entertainment artifice but still existed within a material environment. It was water that lubricated the machine of the desert metropolis. The perpetual rebuilding and reshaping of the strip, the imploding and exploding, presented a constant stream of newness at Las Vegas, a re-rising in the desert. Such newness negated the reality of a tired and decaying city. Constant explosions and rebuilds avoided confrontation with environmental reality. The shiny gold reflective glass of the Wynn casino towers reflected back on to the desert, mirroring the sand.

But the real gamble of Vegas lay outside the casino halls of Wynn, with sprawling suburban growth testing supplies of water at Lake Mead, the city's lifeline. Meanwhile, searches for new water threatened battles with agriculture outside of the metropolis, of 'craps versus crops'.[8]

Late twentieth-century film poignantly situated Los Angeles and Las Vegas as sites of neo-apocalyptic narrative, places of torment and tragedy. *Resident Evil: Extinction* (2007) explored a deserted, dusted and resource-dry future Vegas while, in *The Day After Tomorrow* (2004), a tornado (harbinger of the end) disassembled the iconic Hollywood Sign of Los Angeles. Desert cities worked to environmental deadlines. With Vegas, the loss of water at the great gamblers' oasis threatened a new urban dust bowl. Historically, such a thing had already happened with the cities of the Anasazi, when their irrigation canals failed. At Vegas, and Los Angeles, together with San Francisco, New York and more, environmental collapse seemed not just on the horizon but somehow natural and anticipated (even welcomed, at least in Hollywood film). Rather than a white-city future, the American metropolis appeared doomed.

Hurricane Katrina

The last world's fair in the United States, the final celebration of the white city, suitably occurred in the Orwellian year of 1984. It also took place at New Orleans. A struggling metropolis by the 1980s, New Orleans seemed an apposite endpoint for the fair, the fate of the two aligned once more. The theme of the fair, 'The World of Rivers – Fresh Waters as a Source of Life', hardly suggested a city on the cusp of a roseate future. And the fair itself paled compared to former exhibitions and amusement park competition, most notably Disney's Epcot in Florida that opened in the 1970s with great hullabaloo and offered a remarkable Disneyfied White City. The New Orleans version seemed conceptually and financially bankrupt on opening day.

Rather than a 'white city' of the future, New Orleans was a black city trapped in the past. With racial foundations, it remained a de facto segregated city a hundred years after the end of slavery. African Americans lived in the central, downtown regions. The Lower Ninth Ward was 98 per cent black. Inner-city public-service cutbacks from the 1980s onwards hit residents hard. In 2000, the recorded poverty rate in New Orleans was 28 per cent. Critics charged that structural, institutionalised racism continued to shape the city. Racial issues also intersected with important environmental dynamics. New Orleans was a city surrounded by water, a 'sinking bowl' with black Americans trapped at its centre. Poor environmental design, inner city politics, shifting economics and white flight left African Americans in areas likely to be flooded. The divided city was thus patrolled by socio-environmental dynamics. Along with Las Vegas and Los Angeles, New Orleans was another city in environmental denial, and heading for disaster.

Described as 'one of the most delicate ecosystems on earth', the Gulf Coast of Louisiana was a fragile region even without human interference. The loss of coastal wetlands by urban development meant that there was no organic bulwark or natural defence against storms and hurricanes. At New Orleans, levees designed to protect against flooding also stopped the flow of water and restricted the building of silt, nature's way of forming storm barriers. The basin was drying out. *Time* declared: 'the shriveled Louisiana coastline is dying a slow death at human hands', as if experiencing its own *Silent Spring* or radioactive holocaust. Vulnerable to hurricanes, with its levees ill maintained, and down town neglected, New Orleans seemed to be tempting catastrophe, inviting doomsday.[9]

Some recognised this. In 2001, a Federal Emergency Management Agency (FEMA) report listed three disasters that were widely anticipated: a terrorist attack on New York City; an earthquake in San Francisco; and a hurricane hitting New Orleans. The report accurately predicted two out of three scenarios. In the same year, *Popular Mechanics* magazine carried an article 'New Orleans is Sinking'. *Scientific American* declared 'New Orleans is a disaster waiting to happen'. The *Times Picayune* ran a special on the imperilled southern city, noting 'It's only a matter of time before south Louisiana takes a direct hit from a major hurricane'. *Times* journalists tapped the language of catastrophe culture by employing alarmist phrases such as 'in harm's way' and 'hurricane as nature's ultimate weapon'. In response to the FEMA ranking, Eric Berger for the *Houston Chronicle* described the fast decaying state of New Orleans: development out of control; a deteriorating ecosystem; and the deepening city 'bowl' thanks to pumping technology drying it out. Scholars cited Hurricane Betsy (1956) as historical proof of what might go wrong. An environmental–technological fix in the form of a new reef or floodgate was nowhere near ready. Berger explained 'New Orleans is sinking. And its main buffer from a hurricane, the protective Mississippi River delta, is quickly eroding away, leaving the historic city perilously close to disaster.' The headline ran, 'Keeping its head above water: New Orleans faces doomsday scenario'.[10]

In July 2004, FEMA again sounded the alarm over New Orleans. A 24 July press release read:

> sustained winds of 120 mph, up to 20 inches of rain in parts of southeast Louisiana and storm surge that topped levees in the New Orleans area. More than one million residents evacuated and Hurricane Pam destroyed 500,000–600,000 buildings.

Rated a category 3 hurricane, Pam hit the New Orleans region hard. The 'catastrophic hurricane' quickly led to levees breaking and the city submerged under 10 feet (3 m) of water. Some 270 emergency officials gathered at Baton Rouge for eight days to tackle issues of evacuation, debris, shelter, search and rescue, schooling and medical supplies. Reports came in suggesting a one-year pause before re-entry to the flooded zone, and the rendering of 500,000 Americans homeless. The disaster caused the deaths of 60,000 Americans. But no real hurricane hit. Hurricane Pam was, in fact, a fictional disaster, the imaginative result of a $1 million project to improve disaster planning and emergency response in the Louisiana region. It was the work of Innovative Emergency Management (IEM), a consultancy firm specialising in fictive catastrophes or, as IEM described itself, a 'disaster consulting firm'. Since 1985, IEM had been 'dedicated to keeping people safe – at home, at work, and on the battlefield', and worked with state and federal authorities to 'improve preparedness for hazards ranging from natural disasters to those involving chemical, biological, radiological, and nuclear threats'. With Hurricane Pam, IEM presented local, state and federal authorities with a true 'catastrophe scenario' to manage and work together on. Three films by ADCIRC simulated the visual effects of the hurricane. Officials gathered around a tabletop disaster. Computer screens documented the death toll. According to IEM, 'Participants knew that they were facing a real threat, as articulated in the detailed scenario.' Mock danger, simulated disaster, like a rollercoaster ride at a Disney theme park, engaged the imagination. Pam provided a model of doomsday, a hurricane twin to Doom Town at Nevada Test Site. It gathered together a range of actors awaiting Armageddon. 'Hurricane Pam' provided a dress rehearsal for a real New Orleans doomsday scenario, an eerie dry run for the hurricane ahead.[11]

On 24 August 2005 IEM held a follow-up workshop on Hurricane Pam. Four days later, the actual hurricane hit. The doomsday unfolded on the Louisiana coast. On Sunday, 28 August, Hurricane Katrina headed towards New Orleans with 145-mile-per-hour winds. The National Hurricane Center rated Katrina a 5 maximum storm. City Major Ray Nagin ordered a mandatory emergency evacuation of New Orleans. Out of a total population of 484,000, 80 per cent of residents managed to escape. On Monday, 29 August, winds and storms battered the coastline. As waters surged, city levees began to fail. By Tuesday, 80 per cent of the city had flooded. Fires raged, looting occurred. Overwhelmed by some 25,000 evacuees, and low in supplies, the Louisiana Superdome struggled to support its visitors. The Convention Center opened to cope with extra numbers but similarly strained. Nagin spoke live to CNN, 'a desperate SOS . . . I keep hearing [help] is coming. This is coming. That is coming. My answer is that today is, "BS: where's the beef? . . . They're spinning and people are dying down here."' Federal relief proved slow to arrive, taking until Friday for full rescue and evacuation services to function.

Hurricane Katrina was Hurricane Pam for real. *USA Today* noted how the 'Katrina simulation' proved 'grimly accurate', the 'fictitious storm eerily foreshadowed the havoc wrought'. In reference to Pam's predicted 61,290 dead in a 'catastrophic flood', FEMA director Michael Brown responded 'I pray to God we don't see those numbers'. Fears grew that 'countless corpses lie submerged beneath a toxic gumbo that engulfed the city after the levees gave way'. Emergency responders utilised Pam techniques, including search and rescue employing a 'lily pad' approach. Katrina accurately replicated Pam in terms of debris accumulation and flooding. The IEM noted that environmental consequences were 'eerily echoed in the impact of Hurricane Katrina'. But the hope of organisations working together never happened in the real catastrophe.[12]

The doomsday scenario unfolding with Katrina proved to be a sensational media spectacle across the United States. Live coverage depicted storms attacking the city and complete chaos unfolding in the streets. Nature's fury, a growing evacuation crisis, and the failure of levees to keep Katrina at bay all proved primetime stories. Statistics later emerged of $100 billion property loss, 300,000 buildings gone, two million people displaced, and 1,836 dead. New Orleans had been emptied of life and filled with water. Abandoned residents witnessed the death of their American city. Katrina was labelled the ultimate environmental disaster.

With obliterated buildings, floating carcasses and the stench of death, New Orleans resembled an apocalyptic landscape straight out of Hollywood, a landscape of death documented in the world-wide press. Popular imaginations of doomsday combined with hearsay and delusion. Stories included medicated people dying alone, people left behind in evacuation, stranded and desolate. Rumours spread of a city besieged by looters and overrun by barbarism. Live reportage dramatised the numbers of rapes, deaths, and petty crime at the Superdome. Continual misinformation fuelled the doomsday picture. *Time* commented on how 'Delusion seemed to follow the deluge by a matter of hours'. Delusion and catastrophe became bedfellows of Katrina: popular disaster fiction, public anxiety and the real horrors on the streets of New Orleans fused. This potent doomsday image at New Orleans gathered together a variety of concerns. Catastrophe culture of Katrina featured images of 9/11, environmental collapse, global warming, racial conflict and fears over the end of the American city.[13]

In the popular mindset, the catastrophe culture of Katrina drew on a range of disaster influences. Katrina was compared with other major catastrophes of the American past. ABC likened the exodus of New Orleans residents to 1930s Dust Bowl escapees, both groups running from environmental disaster. The most dominant and powerful reference for disaster proved to be 9/11. The psychological hold of the terrorist attack on the American nation subtly framed Katrina into a 'post 9/11 event'. The 9/11 attack served as the popular reference point and the temperature gauge for all modern disaster. It replaced world war battlefields and Cold War nuclear scenarios as the ground zero for negotiating disaster. It signified the new Geiger counter of catastrophe. The War on Terror hence metamorphosed into a fight against nature at New Orleans. The danger remained something 'outside' the United States, the external threat. One spectator likened the chaos and deprivation of New Orleans to the horrors of downtown Baghdad. One press photographer wrote, 'I think New Orleans had issues before that nobody looked into. This was an ugly, dirty version of 9/11.' Commentators puzzled over the failure of emergency response teams given the rise of the 9/11 State with its comprehensive disaster planning. Post-9/11 America was meant to handle such crises. As scholar Kevin Rozario put it, 'Although the "war on terror" has enabled a massive expansion of the disaster-security state, this did not seem to have made the management of catastrophe any more effective.' Government institutions such as FEMA were found lacking.[14]

The 9/11 attacks also constructed disaster into a narrative of heroes, of brave firefighters and public sacrifice. By comparison, Katrina seemed a disappointing story of criminality. The absence of celebrated heroes in Katrina played off against the hero worship of 9/11. The human story of Katrina seemed, at best, those left behind in the evacuation: the poor, the black, the homeless, and the ill. It amounted to a social embarrassment. The Select Bipartisan Committee to Investigate the Preparation for and Responsibility to Hurricane Katrina labelled the event a 'national failure'. New Orleans resembled a third-world landscape. Katrina emerged as a national shame against the 'glory' of 9/11. When President George W. Bush announced, 'We'll not just rebuild, we'll build higher and better', citizens were reminded of the spectre of 9/11 and the reconstruction of New York City. The *New York Times* responded favourably to the president's words: 'We in New York remember well what it was like for the country to rally around our city in a desperate hour. New York survived and has flourished. New Orleans can too.' But others doubted that New Orleans could emerge victorious from the muddy waters of Katrina.[15]

Scientists situated Katrina within a discourse of meteorology, and part of a long line of hurricanes that battered the American South. Historically, Katrina rated alongside the strongest and most devastating meteorological events, including the Galveston, Texas hurricane of 1900 with its 145-mile-per-hour winds and death toll of around 8,000. Experts placed Katrina within a scientific hurricane-cycle framework, comprehending its meaning by reference to statistics and readings. According to data, a thirty-year period of 'calm' (low activity) led to a new period of 'increased activity' from 1995 onwards. Other hurricanes and affected landscapes came into focus in scientific circles and in the popular press, the seasonal damage to Florida cited as another illustration of nature's force. Commenting on the threat of hurricanes to the Sunshine

State, scholar Ted Steinberg determined that, 'even a passing observer of Florida's landscape can discern the future of the apocalypse'.[16]

Other environmental disasters figured in the Katrina gaze. Analyst Michael Powers positioned Katrina within a disaster framework of human–nature relations, and compared it with the Chicago Fire (1871), San Francisco earthquake (1906), and Dust Bowl (1930s). Powers warned that such disasters 'should be cause for careful consideration of the relationship between human society and naturally occurring events'. Katrina became part of a 'disaster timeline', *Time* magazine placing it alongside a range of other US accidents. Katrina joined a catalogue of catastrophes that included the Schoolhouse blizzard (1888) and the San Francisco earthquake. Scholar Michael Eric Dyson inventively paralleled Hurricane Katrina to the eruption of Vesuvius, with the hit to poor black Americans in New Orleans equivalent to the ash-swamped poorer classes killed in Pompeii. Most commentators posed the possibility of something greater and more powerful on the horizon – a doomsday ahead. The potential for an oil-based environmental disaster in the Gulf seemed one possible scenario, given that Katrina itself rendered five hundred rigs out of action and closed nine refineries. Prior to *Deepwater Horizon*, *Time* commented on the oil region being a 'magnet for hurricanes'. Part of the power of Katrina was the way in which it became seen as a portent of things to come, the eye of a greater storm.[17]

The popular press, scientists and the public also connected Katrina with global warming. Noting the rise of hurricanes in the Atlantic basin, both in numbers and in intensity, Jeffrey Kluger for *Time* pondered 'Is Global Warming fuelling Katrina?' Between 1995 and 1999, a record thirty-three hit the basin. Kluger explained how the warming of oceans contributed to the shift, and that ultimately, 'we have only ourselves – and our global warming ways – to blame'. He feared, 'While the people of New Orleans may not see another hurricane for years, the next one they do see could make even Katrina look mild.' ABC News posited 'Did Global Warming boost Hurricane Katrina's fury?' and claimed that 'If Katrina had struck the Gulf Coast just three decades earlier, the results might have been quite different'. Citing a rise in ocean temperatures (as well as water vapour), ABC charged, 'the finger of blame points quite clearly at global warming'. Such headlines directly linked Hurricane Katrina with concerns over the global environment. As if Katrina lacked impact, the popular media raised the spectre of something worse ahead, and posited a broad trajectory toward environmental doomsday.[18]

The scientific community proved less vocal in making such connections though MIT scholar Kerry Emanuel's data on rainfall, water vapour and sea temperatures were widely referenced. Emanuel discovered an increase in storm intensity due to global effects. Research at MIT on tropical cyclones supported the contention that the intensity of storms had increased in the post-war period, citing warming as the likely trigger. Climatologist Kevin Trenberth noted how, at New Orleans, 'levees were designed in times that didn't take into account the enhancement or global warming factors, let alone Category 5 storms'. The scientific community, however, struggled with the exact contribution of global warming to Hurricane Katrina. As the *New Scientist* highlighted, how could anyone prove a direct causality between a gradual global shift and a specific natural disaster? At best, if more 'extreme events' occurred, the finger pointed to global warming or, as the magazine put it, 'if sixes keep coming up more often than the other numbers, you know the dice is [*sic*] loaded'. Humanity seemed to be playing some sort of loaded doomsday roulette.[19]

The link between global warming and Hurricane Katrina proved even more challenging to substantiate in the courtroom. A group of fourteen Mississippi residents affected by Hurricane Katrina attempted to sue a group of energy and chemical corporations for their contribution to global warming, and thus Katrina. 'The plaintiffs allege that defendants' operation of energy, fossil fuels, and chemical industries in the United States caused the emission of greenhouse gasses that contributed to global warming,' read the court document. After the United States District Court rejected the claim, the Fifth US Circuit Court of Appeals in October 2009 ruled that the plaintiffs did have standing and grounds to sue. In January 2011, however, the United States Supreme Court rejected the petition and dismissed *Ned Comer et al.* v. *Murphy Oil USA et al.*[20]

Katrina also featured in the global warming blockbuster documentary, Al Gore's *An Inconvenient Truth* (2006). Drawing on Gore's background as an environmental campaigner (in 1992, he wrote *Earth in the Balance*), *An Inconvenient Truth* documented a world of melting ice caps and death by carbon dioxide. The film projected a human-caused catastrophe, an environmental doomsday ahead. The reception to Gore's opus proved significant. The sixth highest-grossing documentary of all time and two Academy awards testified to its power. NASA climatologist James H. Hansen claimed that, with *An Inconvenient Truth*, 'Al Gore may have done for global warming what *Silent Spring* did for pesticides'. *The New Yorker*'s David Remnick noted that while Gore's documentary was 'not the most entertaining film of the year . . . it might be the most important', by providing a 'brilliantly lucid, often riveting attempt to warn Americans off our hellbent path to global suicide'. Katrina featured in the film as a key emotional image of disaster. 'How, in God's name, could that happen here?' begged Gore. Katrina serviced as a warning sign, a symbol of greater devastation, a harbinger of doom.[21]

Others recognised in Katrina a fundamental message over the relationship between humans and the environment. Responsible for keeping Lake Pontchartrain north of the city and the Mississippi from closing in from the south, the 350-mile (560-km) system of levees of New Orleans proved to be a vital divider between nature and civilisation. The levees kept the city safe by preventing the low parts of New Orleans from being flooded. But engineers failed to keep nature at bay. Nature showed the upper hand. Environmental engineer Rafael Bras mused, 'You'll never be able to control nature. The best way is to understand how nature works and make it work in our favor.' The Katrina doomsday image: a ruined city; widespread dereliction and decimation; a world of flattened landscapes and a simple demonstration of nature's power. The hurricane deconstructed human achievements, disassembled the city space. Structures were reduced to their elemental states. Houses became floating planks of timber and brick piles. Nature washed away civilisation, broke down the metropolis. In the face of the flood, the American metropolis collapsed.[22]

And while civilisation fell, a watery wilderness took hold. Playing on folkloric wilderness fears, journalists told of infested dirty waters, of sharks, snakes and alligators patrolling the post-apocalyptic New Orleans. Photographs depicted rooftops just above the waterline, the last vestiges of culture surrounded by a sea of brown. One reporter described how 'the waters became incredibly foul, a vile brew of gasoline, sludge, snakes and canal rats, stinking of sewage and decaying bodies'. Fears of drowning in contaminated water and of deadly creatures in the sea proliferated. Ted Steinberg compared Katrina to the Dutch punitive 'drowning cell' of the sixteenth and seventeenth centuries, a cage designed

Figure 6.1 A US Army High Mobility Multipurpose Wheeled Vehicle traverses through floodwater surrounding the Superdome in New Orleans, Louisiana, September 2005 (US Air Force photograph by Staff Sgt Jacob N. Bailey; Courtesy of US Army).

to torture and trap its inmates. A small number in New Orleans experienced the torture first hand, one inmate locked in a cell in the New Orleans Parish Prison, trapped as the water headed in (later referenced in the film *Bad Lieutenant* [2009]). With images of a flooded city, of dead bodies floating on the water, scenes took on an almost mythological element, that of the biblical flood.[23]

Commentators pondered nature's specific contribution to this doomsday. Was nature out of our control? Portraying American citizens (and, more importantly, government) as innocent victims, Dick Cheney stated, 'You've got to recognize the severity of what Mother Nature did to us here'. Nature's fury destroyed New Orleans. Americans seemed powerless before the storm. The creations of humankind and technology suddenly became small and insignificant during Katrina, dwarfed by the hurricane. Weighing some 13,000 tons, the Chemul oil rig surfed the water before hitting Cochrane–Africatown USA Bridge, testimony to the power of hurricane forces. Like the disaster movie *2012*, nature hit hard and both humankind and technology faltered. Wild nature seemed capable of ending civilisation. Katrina threatened culture, place, and past. An environmental disaster threatened to wipe out a city. As *Time* related, 'amid the flooded streets of New Orleans, the memory of good times past is obscured'. Katrina marked the fall of Babel, dereliction in the Garden of Eden.[24]

The doomsday narrative further drew on a classic fear of catastrophe culture: the fall into barbarism that corresponded with disaster. The levees demarcated the line between civilisation and savagery. When they fell, the city fell. Looting incidents were verbatim proof of new dangers. The fleeing of police (around one-third disappeared) exacerbated the scenario. The city fell into chaos for a short time. *Time* called it a 'zone of anarchy'. Within this narrative, the Superdome became a place of last resort, a last stand of civilisation, akin to a post-nuclear bunker. Outside the dome, rescue efforts grew. Helicopters swept in to repair dam levees and rescue children. Revisiting the horrors of Vietnam, Army Corps of Engineers spokesperson Susan Jean Jackson declared: 'It's like Apocalypse Now'.[25]

Katrina and racial disaster

Whatever the exact contribution of nature or global warming to Katrina, it was clear that American actions contributed to the event, and exacerbated the scale of disaster. Most commentators came to one conclusion: America could not handle a large-scale environmental disaster. A long list of errors and oversights at Katrina were commonly cited. They included: an Amtrak train being offered to the city for evacuation, then rejected; a Louisiana health official turning down help from the Department of Health; the failure of the emergency office at New Orleans; a poor response to the disaster by the National Guard; a flawed bus evacuation; and key failures in emergency governance. While the reversal of three highways facilitated an evacuation of the 'mobile', 20 per cent of residents were left behind over the initial weekend as Katrina grew in power. Emergency housing, in the form of poor (even toxic) trailers, failed to help matters. Housing expert Sheila Crowley nicknamed it 'a disaster of a disaster response'. FEMA and the Department of Homeland Security were widely criticised. Their failures included a slow response time, communication difficulties, and poor management of the Superdome. Rather than being successfully evacuated, thousands found themselves stranded for days in the middle of New Orleans. Volunteer efforts replaced government-organised rescue, while the USS *Bataan* rested idly off the coast, 600 beds ready for use, throughout the disaster. Foreign emergency supplies were put on hold. A memo from Mayor Nagin to FEMA highlighted the needs of the city, and the reality that most had not been met. In response, FEMA director Michael Brown (who Steinberg colourfully claimed 'knew more about horses than hurricanes') later resigned. Katrina produced a momentary crisis of confidence in government. Katrina also represented a blow to the esteem and reputation of a nation. The sense that America had failed to protect its citizens in an environmental disaster worried many. Katrina encouraged the growth of catastrophe culture and a disaster-framed society. One Pew Research Center survey in 2010 showed that, five years after Katrina, 57 per cent of Americans believed that the nation was no better prepared for hurricanes and other natural disasters.[26]

The claim that the United States dealt poorly with Katrina also involved significant racial commentary. The media broadcast the shocking reality that environmental disaster failed to hit citizens equally. Images showed that the overwhelming number of Americans who suffered were black. Katrina provided a story of the survival of the richest (white Americans) with the lowest earners, poor and sick (predominantly black Americans) left abandoned. In that sense, Katrina was a black doomsday, a disaster scenario reserved for people of colour. The loss of New Orleans was also a loss of black culture, of jazz, Louis Armstrong and Wynton Marsalis's America's soul kitchen. The death of celebrated songwriter Fats Domino was rumoured during the disaster (Domino was rescued by the Coast Guard). Katrina thus seemed different from typical apocalyptic imaginations propagated by white America and Hollywood.

The racial past and present of New Orleans determined the Katrina doomsday picture. Historic geography moulded the disaster, cut the paths for specific lines of damage. Slavery made the city, shaped the place, and influenced the disaster. The black centre of New Orleans, hit hardest by the hurricane, was the product of hundreds of years of slavery. Dating back to 1718, Creole, Spanish, French and Anglo influences all fashioned New Orleans. It served as a hub of agricultural trade in the antebellum South, and entertained 'the busiest slave market'. During the nineteenth century, well-to-do families settled in city areas of low risk and high reputation (such as the Garden District) while, from the 1860s, freed slaves sought homes in the 'least desirable' parts of town. Transport and the opening of new land (often exclusively signed over to whites) facilitated movement out of New Orleans. From the 1960s on, white flight to suburban outskirts, along with Housing Authority projects inside, left downtown New Orleans overwhelmingly black. The central population shifted from 37 to 67 per cent black. Richer, higher-level suburbs were typically white, poorer, lower areas black. Inside the city, public disinvestment and welfare cuts worsened living standards. Outside the city, along the Mississippi River, a range of chemical plants and toxic dumps (nicknamed Cancer Alley) was situated in a black community. This socio-environmental hierarchy of whites on top and blacks at the bottom melded racism with topography. Press photographer Thomas Dworzak labelled it the 'geography of racism'. New Orleans resembled an apartheid city, a realm of residential segregation. Race scholars talked of 'the myriad ways that converged long before Katrina made landfall to make New Orleans ripe for a disaster that would hit the city's black residents the hardest'. The shape of natural disaster seemed eerily predetermined. [27]

When the storm hit, it followed those socio-environmental contour lines. Water flooded the black areas that lacked sufficient levee protection while white areas, such as the French Quarter and Garden District, coped better. Structural racism intertwined with environmental disaster. Politician Jesse Jackson labelled the overwhelmed New Orleans Convention Center as like 'the hull of a slave ship' during the panic. Katrina took on the countenance of a neo-slave landscape. For Dyson, Katrina was fundamentally 'a Southern racial narrative being performed before a national and global audience'. Katrina was cited as yet another example of how 'black grief and pain have been ignored throughout the nation's history'. One interviewee cried live on television, 'Business as usual for this country. When is the last time America lifted a finger to do right by a Black man? We were worth more when we were slaves.'[28]

Media highlighted, as well as played into, racial arguments. During coverage of the event, journalists showed that racism was still alive in the United States. But they also relied on a tired language of race. Some reporters grouped black Americans as 'they' (the other), 'victims' rendered powerless by the storm. Rather than escaping the carnage, black Americans became 'refugees' as if without nation. Captions for two photographs highlighted the language of race: one featured white Americans finding free groceries while another image declared blacks 'looters'. Rumours of rampaging black Americans in New Orleans regurgitated old myths of the 'dangerous black man': crime stereotypes interfaced with slave stereotypes perpetuating racist imagery. As Hillary Potter related, journalists offered, 'rumors "othering" Blacks as somehow "less worthy" due their criminal, violent, or welfare-driven natures'.[29]

Meanwhile, the ineptitude of the federal relief effort led to wide-ranging accusations of racism in government circles. Many felt that, if white Americans had been stranded, the rescue efforts would have proved far quicker. The popular face of mass media, celebrities cast their eyes over the evacuation of New Orleans. Hollywood actor Colin Farrell charged, 'If it was a bunch of white people on roofs in the Hamptons I don't have any f******* doubt there would have been every single helicopter, every plane, every single means that the government has to help these people.' Jesse Jackson cried, 'Bush has not shown that he cares for civil rights or cares for the interests of black people.' Live on a NBC telethon for Katrina victims, 2 September 2005, rapper Kanye West vocalised his concerns. With comedian Mike Myers nervously keeping to his autocue, West deviated from script, stating to the camera, 'George Bush doesn't care about black people'. The delivery had power thanks to the combination of celebrity status, emotion and vitriol. It also reflected mass doubts over the policies of the Bush administration. Some charged that the administration had hardly looked after black Americans before Katrina struck, that social welfare cuts had already left blacks destitute and abandoned. Katrina thus signified the final cut. President Obama later criticised the 'passive indifference' of FEMA and Homeland Security.[30]

Thus, for some, the function of the storm over Katrina was to highlight the greater storm over race relations in America. Trapped by both their historic predicament and the disaster at hand, black Americans in New Orleans were rendered powerless over their destiny, their freedom curtailed. For Potter, this was the true 'catastrophic disaster of epic proportions': the systematic control and disenfranchisement of black Americans. Pointing to the effect of racism and poverty on the city, some scholars even declared, 'There is no such thing as a natural disaster'. To call it natural was a failure to acknowledge the racial dynamics at play. Katrina was instead identified as a 'social disaster', the inevitable end product of years of neglect. Arguably, there was not even an environmental component left in the story. Katrina was something that could have been avoided with the right attitudes and right policies.[31]

Katrina also provided a lens on wider social problems in America. It exposed the harsh realities for many residing in the United States, especially the 36-million-strong underbelly. An environmental disaster became a vehicle for criticising American society. Rozario claimed that Katrina helped question the American wealthy and the American way. *Time* magazine declared Katrina 'The Storm that changed America'. But, within a few years, sympathy for the black poor seemed far less apparent. American supercorporations, such as Walmart, meanwhile benefited most from the city rebuild.[32]

Death of the city

My whole life, my whole world crashed. For everyone, not just me.

Robert Fontaine, New Orleans resident

Time magazine portrayed New Orleans as a mythological, almost Atlantis-like place before Katrina. It venerated the city for its jazz, architecture and cultural diversity. The magazine declared, 'New Orleans is a true American melting pot'. In just a few days, this Utopia was transformed into a watery hellhole. A beautiful coastline was replaced with 90,000 square miles (223,000 sq. km) of debris, waste and dereliction. The National Resources Defense Council (NRDC) classified Katrina, 'perhaps the single worst environmental catastrophe to befall the United States as a result of a natural disaster', and an 'indelible mark' on the landscape. Whole districts, freeways and cars were swamped and lost. The scale of environmental damage included 7 million gallons (26.5 million litres) of oil spilt, 575 chemical spills reported and four hazardous waste sites hit. A heady cocktail of toxins, petrochemicals and sewage was released. The 'overpowering smell' of sewage and decay hit every visitor. As one reporter related, 'The city smells dead'.[33]

Katrina represented the fall of the American city at the beginning of the twenty-first century and signified the loss of American greatness. Katrina wiped clean the culture, heritage, and identity

Figure 6.2 'Abandoned Six Flags amusement park in New Orleans wrecked by Hurricane Katrina' (courtesy of Keoni Cabral).

of place. Narratives imparted the huge cost. The *Washington Post* described the 'empty city' of disaster, a city poisoned. 'On such grand boulevards as St. Charles and Napoleon avenues, the foliage that drapes the majestic oaks and magnolia trees is suddenly turning brown. Birdsong has largely disappeared, replaced with the whine of boat engines and the shouts of rescuers seeking survivors,' the paper related. Katrina's New Orleans resembled the 'Fable for Tomorrow' town of Carson's *Silent Spring*, an idealised place, now polluted, dead and gone. In work on 'Katrina's environmental fallout', the NRDC painted a picture akin to a bombed-out test site: of whole districts destroyed, with 'the hard-hit areas ... largely ghost towns'.[34]

Others saw an already defunct city in its final stage of decay. Crime, governmental mismanagement, poor economics and poverty had killed New Orleans before Katrina. One newspaper labelled it 'The City that care forgot'. Meanwhile deficient urban design had 'prepared' it for environmental Armageddon, invited doomsday. *Newsweek* described the historic development of New Orleans as the 'granddaddy of all environmentally misguided plans'. Natural limits had been ignored. The hurricane wiped the slate clean.[35]

On a wider, symbolic level, the loss of New Orleans embodied the demise of the great American city. The hurricane dealt a death blow to a certain idea of the metropolis. The destruction of the city reflected the broader decline of American urban society. 'America is once more plunged into a snake pit of anarchy, death, looting, raping, marauding thugs, suffering innocents, a shattered infrastructure, a gutted police force, insufficient troop levels and criminally negligent government planning' wrote Maureen Dowd for the *New York Times*. 'It attacked not just a landscape, not just homes and buildings: it also attacked an idea of a city that is deeply rooted in America's history and culture', related *Time*. The dominant image of Katrina's New Orleans was one of environmental–social ruin and a realm of suffering and death. The disaster of Katrina hit New Orleans hard, and some feared a knock-on effect, with the collapse of other structures and cityscapes.[36]

Could New Orleans, and with it, the modern city, be resurrected? On 15 September 2005, President Bush foretold, 'there is no way to imagine America without New Orleans, and this great city will rise again'. Scholars pointed to the opportunity presented by Katrina to start anew with New Orleans. With the 'slate' swept clean, came the opportunity to rectify linkages between race, poverty and a toxic environment.[37]

But hyperbole failed to translate into the speedy rebirth of New Orleans: a new Atlantis needed time. Given all the damage, it was hard to imagine a thriving metropolis again. Some 150,000 to 200,000 motor cars needed to be removed; debris and water required attention; and the canals and levees urgently demanded repair. Checks were needed on the saltwater contamination of freshwater lakes. The NRDC reported significant health risk in the region because of the presence of DDT, petroleum, lead and other toxins in water, soil and debris. Post-Katrina New Orleans was a pollutant hotspot. The *Post* headlined, 'Katrina takes environmental toll', detailing how, 'The dank and putrid floodwaters choking this once-gracious city are so poisoned with gasoline, industrial chemicals, feces and other contaminants'. The newspaper predicted that contamination, infection and disease would linger for years to come. For sociologist Robert Bullard, Hurricane Katrina, combined with environmental racism, created a 'potential toxic dump'. Journalists questioned whether the revival of the dead and flooded city of New Orleans was even possible. Could the great American city survive a doomsday, could it weather the incremental hurricanes? In a December 2005 article, entitled 'Death of an American City', the *New York Times* pondered such issues. It critiqued reconstruction efforts and clean-up operations. It raised the spectre of New Orleans becoming a modern-day ghost town, or a defunct and closed amusement park.[38]

In the 1920s, the Chicago School pioneered new mapping techniques for sociology. Robert E. Park and Ernest Burgess designed a map of Chicago based around concentric circles and zones of living. The approach offered the city as a kind of organism, a living, breathing social world. In the 1990s, Mike Davis colourfully remapped the city of Los Angeles employing the same approach. Davis drew a city of ecological and social hell. A map of New Orleans might show similar things at work as in Los Angeles, with the Louisiana city depicted as a swamped, decaying organism.[39]

Katrina showed a real environmental disaster hitting a real city but Hollywood was also doing similar work. In *The Day After Tomorrow* (2004), doomsday was tied to the fall of the great American city of New York. Detailing the end of the world, the film illustrated disaster with the fall of the Big Apple. Set out in the Hollywood blockbuster, post 9/11, neo-Cold War anxieties over the loss of New York City to terrorists combined with disquiet over global warming. *The Day After Tomorrow* updated doomsday fears, and replaced Cold War anxieties over nuclear attack and Carson's chemical fallout with fresh concerns over terror and deluge. Global warming became the new nuclear fear. *The Day After Tomorrow* thus functioned as a revision of *The Day After* (1983), a film that documented the demise of the American city, in that case Kansas City, to radioactive fallout. From the day after to the day after tomorrow, the American city remained the ground zero of doomsday and the centrepiece of eco-problems.

Notes

1. Even at the Chicago Expo, Louisiana's contribution to the state exhibits hall was a recreation of a Creole mansion. With 'ancient spinning wheels and looms of an industry of olden days', the state exhibit harkened back to the colonial era. Louisiana presented itself as aristocratic, elitist and backward. See Hubert Howe Bancroft, *The Book of the Fair* (1893).
2. Upton Sinclair, *The Jungle* (1906), pp. 128–9.

3. Jacob Riis, *How the Other Half Lives* (1890), pp. 2–3.

4. Lewis Mumford, *The City in History* (San Diego: Harcourt Brace, 1961), p. 603; Jack London, "Story of an Eye-Witness," *Collier's* 5 May 1906.

5. Paul Ehrlich, *The Population Bomb* (1968), p. 1; 'Briton Criticizes Prophets of Doom', *The Rock Hill Herald*, 12 May 1971; Fairfield Osborn, *Our Plundered Planet* (Boston: Little Brown, 1948), pp. 194, 37; Mumford (1961), p. 602.

6. Paul Simmons, "Patterns and Trends in Overcrowded Housing," *Redefining Urban and Suburban America* (Washington, DC: Brookings Institution, 2005).

7. Mumford (1961), p. 545.

8. 'Las Vegas Could Run Dry by 2021', *Environmental Graffiti*, 14 February 2008; 'Vegas Heading for Dry Future', 29 July 2005, BBC News.

9. *Time, Hurricane Katrina: The Storm that Changed America* (2005), p. 74.

10. FEMA Press Release, 24 July 2004; Jim Wilson, 'New Orleans is Sinking', *Popular Mechanics*, 11 September 2001; Mark Fischetti, 'Drowning New Orleans', *Scientific American*, October 2001; 'Special Report: Washing Away', *The Times Pacuyne*, 23–7 June 2002; *Houston Chronicle*, 1 December 2001.

11. FEMA Press Release, 24 July 2004; IEM, 'Preparing for a Catastrophe', statement before Senate Homeland Security, 24 January 2006. With 'Hurricane Pam', officials worked collectively to produce a detailed response plan. The IEM's 'Preparing for a Catastrophe' document was one result. FEMA regional director Ron Castleman enthused, 'We made great progress this week in our preparedness efforts'. Funding problems for project, however, meant that it was never completed.

12. 'Government simulation predicted 61,290 deaths', *USA Today*, 9 September 2005; IEM, 'Preparing for Catastrophe'.

13. *Time* (2005), p. 55.

14. 'Lessons from the Dust Bowl for Hurricane Survivors', *ABC News*, 23 September 2005; *Time* (2005), pp. 111, 59; Kevin Rozario, *The Culture of Calamity: Disaster and the Making of Modern America* (Chicago: University of Chicago Press, 2007), p. 212.

15. Final Report of the Select Bipartisan Committee to Investigate the Preparation for the Response to Hurricane Katrina (2006), p. x; 'President Discusses Hurricane Relief in Address to Nation', 15 September 2005, White House Press Release; 'Death of an American City', *New York Times*, 11 December 2005.

16. Ted Steinberg, *Acts of God: The Unnatural History of Natural Disaster in America* (Oxford: Oxford University Press, 2006 [2000]), p. 206.

17. Michael Powers, 'A Matter of Choice', Chester Hartman and Gregory Squires, eds, *There is No Such Thing as a Natural Disaster: Race, Class and Hurricane Katrina* (New York: Routledge, 2006), pp. 25–6; *Time* (2005), pp. 74–5; Michael Eric Dyson, *Come Hell or High Water: Hurricane Katrina and the Color of Disaster* (New York: Basic, 2005), p. x.

18. Jeffrey Kluger, 'Is Global Warming Fueling Katrina', *Time*, 29 August 2005; 'Did Global Warming Boost Katrina's Fury?', *ABC News*, 14 September 2005. Also see 'Katrina's real name', *The Boston Globe*, 30 August 2005, Joseph Verrengia, 'Katrina reignites global warming debate', *USA Today*, 1 September 2005.

19. *ABC News*, 14 September 2005; 'Climate Myths: Hurricane Katrina was caused by global warming', *New Scientist*, 16 May 2007.

20. 'Did "Big Oil" cause Hurricane Katrina', *American Infrastructure*, 3 September 2010.

21. Jim Hansen, 'The Threat to the Planet', *New York Review of Books*, 13 July 2006; David Remnick, 'Ozone Man', *The New Yorker*, 24 April 2006.

22. Rafael Bras, *New York Times*, 6 September 2005.

23. *Time* (2005), p. 7; Steinberg (2006), p. 197.

24. *Time* (2005), p. v.

25. *Time* (2005), pp. 16, 25.

26. Sheila Crowley, 'Where is Home? Housing for Low-Income People After the 2005 Hurricanes', Chester Hartman et al., *There is No Such Thing as a Natural Disaster: Race, Class and Hurricane Katrina* (New York: Routledge, 2006), p. 129; Steinberg

(2006), p. 199; Pew, 'Five Years After Katrina, Most Say Nation is Not Better Prepared', 26 August 2010, press release.

27. Richard Campanella, 'An Ethnic Geography of New Orleans', *Journal of American History* 94 (December 2007), pp. 704–15; *Time* (2005), p. 60; John Powell et al., 'Towards a Transformative View of Race: The Crisis and Opportunity of Katrina', in Hartman (2006), p. 64. Geographer Richard Campanella showed that whites did move to low-lying areas, such as Lakeview, and some black Americans resided on riverfront locations, suggesting perhaps all parties faced the environmental threat.

28. See Arnold Hirsch and A. Lee Levert, 'The Katrina Conspiracies: The Problem of Trust in Rebuilding an American City', *Journal of Urban History* 35/2 (January 2009); Dyson (2005), pp. 21, 24.

29. See Tania Ralli, 'Who's a Looter? In Storm's Aftermath, Pictures Kick Up a Different Kind of Tempest', *New York Times*, 5 September 2005; Michelle Miles and Duke Austin, 'The Color(s) of Crisis', in Hillary Potter, ed., *Racing the Storm* (Lanham: Lexington, 2007), p. 39.

30. Dyson (2005), p. 20.

31. Potter (2007), p. 3; Hartman (2005), p. 3.

32. *Time* (2005), title; Rozario (2007).

33. *Time* (2005), pp. vi, 112; NRDC, 'The Environmental Effects of Hurricane Katrina', report, 6 October 2005.

34. *Washington Post*, 24 October 2005; NRDC (2005).

35. Nicole Gelinas, 'Houston's Noble Experiment', *City Journal* (spring 2006); Anna Quindlen, 'Don't Mess with Mother', *Newsweek* / MSNBC, 19 September 2005.

36. Potter (2007), p. 39; *Time* (2005), pp. 32–3.

37. 'President Discusses Hurricane Relief in Address to Nation', 15 September 2005, White House Press Release; *Time* (2005), p. 116.

38. 'Pollution to Linger for Years', *Washington Post*, 7 September 2005; *NYT*, 11 December 2005.

39. See Mike Davis, *Ecology of Fear: Los Angeles and the Imagination of Disaster* (Basingstoke: Picador, 1998), pp. 364–5.

CHAPTER SEVEN

Disney/Disnature and the End of the Organic

When Italian philosopher Umberto Eco travelled the United States in the 1970s, he visited two versions of New Orleans. One constructed over hundreds of years by white colonists and black slaves. The other constructed in a few months by Walt Disney Imagineers. Part of the entertainment landscape of California's Disneyland, the hyperreal, cartoonised New Orleans impressed Eco far more than the authentic, historic city of the South. Seeing animatronic alligators on the Disney Mississippi but no live specimens near New Orleans, Eco imparted: 'Disneyland tells us that technology can give us more reality than nature can'.[1]

Americans have always manipulated and simulated nature. Even national parks, venerated 'wilderness landscapes', amounted to spectacularly constructed nature experiences in the early 1900s, marked by predator-control campaigns, superabundant prey numbers and elaborate tourist provisions. Historic simulations of nature include Carl Hagenbeck's zoo dioramas of the 1890s and 'nature faker' fiction by William J. Long and Ernest Thompson Seton, authors who granted their fictive animals decidedly human characteristics.

In the late twentieth century, the scale of this simulation shifted a gear. Scientists played with life on a fundamental level. Genetic engineering produced modified crops and Dolly the Sheep. In self-contained laboratories, white-coated researchers tampered with the building blocks of the world outside. Divorced from any tangible link with the great outdoors, computer technology created cyborgs, artificial life, and digital/virtual nature. Rachel Carson's fears that humanity might alter nature on a primary level were realised. Technology began to produce a new post-nature reality.[2]

At the same time, everyday Americans proved to be less connected with the outdoors. This owed much to the rise of cityscapes but also to the proliferation of media and technology. People liked their televisual nature, their plastic plants and their Disney New Orleans alligators. Global pollution sullied pure nature, cast a smog cloud across the landscape, leaving nothing untouched. In *The End of Nature* (1990), Bill McKibben described how the creosote bush had survived for thousands of years, but no more. Everything seemed contaminated. McKibben raised the spectre of no more wilderness or independent life, and presented perhaps the ultimate environmental doomsday scenario.[3]

An exchange certainly seemed to be in evidence: the rise of artificial life concomitant with the fall of material wilderness. With the downward spiral of American nature and the end of 'pure' wilderness, came the rise of artificial nature and the technological facsimile. The meaning of nature shifted from its utility in the nineteenth century as a frontier resource and romantic emblem to its provision of light entertainment on television and Internet in the twentieth century.

This final chapter is about the artificial world of Walt Disney. It explores the process of copying, mimicking and redefining nature through the Disney canon. It charts the rise of Disneynature (or Disnature) through Mickey Mouse, Bambi, and True-Life Adventures, and explores artificial nature as found in Disney parkscapes. It equally ponders the demise of 'real' nature that accompanies the emergence of Disnature. Ultimately, the facsimile project forwarded by the Disney kingdom spells the end of organic nature. Cartoon mouse eats mouse.

DisneyNature

DisneyNature:
1. Movie label belonging to The Walt Disney Company, announced 21 April 2008, focusing on nature documentaries (Brand);
2. Nature as defined and given form by the Disney corporation. Form of anti/post-nature. (Process) Abbreviation: Disnature.

Along with protest action and community gatherings, Earth Days since 21 April 1970 have often entailed some degree of corporate involvement. Compiled for the first Earth Day, the *Environmental Handbook* warned against eco-phonies (particularly corporate giants, such as BP) hijacking the green movement, attempting to promote (or 'greenwash') their products in the process. Could activists competently distinguish between greenwashing and green thinking, or reject corporate propaganda while welcoming environmental capitalism? On 21 April 2008, Walt Disney Company announced a new film stable to produce nature documentaries for the big screen. 'Disneynature', headed by Jean-François Camilleri, and based in France, promised live action wildlife and nature films, and equated to a thorough updating of Disney's own True-Life Adventure series of the 1950s. The first movie in the American market, released on Earth Day 2009, was appropriately titled *Earth*. It took $106 million. *Crimson Wing: Mystery of the Flamingos* followed a year later, with *Oceans* unveiled at the 2010 Earth Day.[4]

'This Earth Day we can celebrate the Earth simply by entering a movie theater and taking a seat!' read one billboard for *Oceans*. The perfect realisation of 'armchair environmentalism', Americans did their bit by watching a film, comfortable in knowing that their cinema ticket committed Disneynature to donate a small percentage of profit to eco-causes. The idea of celebrating Earth Day with Disney tied the media giant with modern environmental consciousness. Treating Earth Day like a Superbowl half-time show, the Disney company latched on to nature's annual calendar event. Planet Earth received a new corporate sponsor.

A Disneyfied Earth Day seemed appropriate to a world dominated by media giants and entertainment culture. For decades the premier force in shaping child (and adult) ideas about nature and animals, the organisation sported a long history in environmental education. But the true environmental record of Disney proved ambiguous. Land clearance for parks, contributions to pollution and dubious projects, such as Mineral King ski resort, tarnished the Disney eco-brand.

And the question remained: what exactly did Disneynature mean? The logo of Disneynature, a giant ice sculpture of Disney's sleeping castle amounted to a naturalistic rendering of the classic cartoon fairytale. Since the 1920s, Disney had promoted its own version of nature: anthropomorphic, facile and American. The caption 'Warning: Film Contains Natural Animal Behavior', in new Disneynature productions shocked some viewers for it suggested unexpected authenticity in the Disney universe. Did the cultural dominance of Disney's reel nature threaten the end of real nature or, at the very least, mark a shift in how Americans understood the natural world? Was the global corporation more dangerous for its physical environmental damage, or for the way it polluted the way people conceived the organic world? Arguably, Disneynature (or Disnature as process) threatened the heart of nature, our understanding of life around us. It serviced a cultural and environmental disconnect and a subtle and invasive doomsday scenario.

The Disney way

Formed in October 1923 by brothers Roy and Walt, the Disney Brothers Cartoon Studio first pioneered animation techniques in a studio in Hollywood, California. Disney won favour for its mascot Mickey Mouse with a series of dynamic productions, beginning with *Steamboat Willie* (1928). Exploiting shrewd marketing opportunities (such as merchandising characters) and new technology (such as Technicolor), Disney prospered in the depression era. Along with Mickey Mouse cartoons, the feature film *Snow White and the Seven Dwarfs* (1937) positioned Disney as a key innovator in American entertainment by the 1940s. During World War II, the corporation produced propaganda films for the United States government and military. *Der Fuehrer's Face* (1942) depicted Donald Duck experiencing a mental breakdown while manufacturing artillery shells for Nazi Germany. Expansion into television, wildlife documentaries, live action films, and theme parks by the 1950s cemented Disney as the premier entertainer in a global market. By 1954, one third of the world population had seen a Disney film. In 1971, Disney opened Walt Disney World in Florida. The death of both Roy and Walt, a lack of innovation, and a hostile takeover bid in 1984 followed. During the 1980s, Disney struggled to make hit animated features. Beginning with *The Little Mermaid* (1989), however, the company prospered under Michael Eisner's direction. *The Lion King* (1994), a *Jungle Book* for the 1990s, made Disney $783 million worldwide. Disney partnered with Pixar, and entered the computer-generated imagery (CGI) market with *Toy Story* (1995). Diversification into more mature films (with the purchase of Miramax in 1993) and the acquisition of ABC strengthened the Disney fold. Disney Stores opened, as did Disneyland Paris (1992). Film success continued with *Finding Nemo* (2003) and *Wall-E* (2008). Entering the twenty-first century, Disney maintained a diverse portfolio, with particular success in the teen television market, evidenced in *High School Musical* (2006).

The goliath media status of the Walt Disney Company proved remarkable given its humble origins. As Walt Disney reminded, 'It all started with a mouse'. The success of Disney owed much to ambitious marketing, consistent branding, technological innovation, and a synthesis of entertainment and education. The input of Walt Disney, along with a high level of control over Disney production lines, also contributed. Drawing on sociologist George Ritzer's work on McDonaldization (1993), Alan Bryman pointed to four specific 'tools' employed by Disney to good effect, especially in terms of its theme park business. For Bryman, Disney relied on theming (the creation of distinct and immersive worlds), hybrid consumption (tying different consumer experiences together), merchandising (of Disney characters) and performative labour (how staff entertained their customers, for example). This operating system equally applied outside the world of Disney, to the extent that, to Bryman, 'modern society is increasingly taking on the characteristics of the Disney theme parks'.[5]

Most scholars agreed that Disney exacted significant influence over American culture and values. Writing in 1973, Christopher Finch declared, 'Disney is a primary force in the expression and formation of American mass consciousness'. Elizabeth Bell claimed, 'it's logo and characters have become almost synonymous with the very notion of American popular culture'. Alongside Coca-Cola and McDonald's, Disney ranked as a paragon of American consumer culture. The Americanism of Disney rested in its patriotic content, its common referents to the American dream, its capitalist zeal, and its innovative flair. The self-contained 'Disney universe' promoted traditional family values and morals, a Protestant work ethic and individualism. Disney identified with conservative, middle-class, white America. Cartoons also featured childlike and simplistic narratives spinning good and evil dichotomies. Witnessed abroad, this 'Disney-America' translated as cultural imperialism.[6]

Critics voiced significant concern over both the content and the impact of the Disney message, especially on childhood. Walt Disney was critiqued for his fervent anti-Communism and lambasted as a racist and an imperialist. Henry Giroux warned against taking Disney at face value, that our 'recognition of the pleasure that Disney provides should not blind us to the realization that Disney is about more than entertainment'. He admonished the 'pretense of innocence' lurking in every Disney product. Bell claimed Disney actively altered the way we see the world. Its themed lands and manufactured stories employed 'a notion of innocence that aggressively rewrites the historical and collective identity of the American past'. Bell declared, 'Disney's trademarked innocence operates on a systematic sanitization of violence, sexuality, and political struggle.' The anodyne, superficial, and childish elements of Disney offended high-culture spectators. Disney inspired vilification alongside adoration.[7]

The actual 'world' or 'universe' of Disney resided in a hotchpotch collection of ideas and concepts. Inspiration came from European folklore, American history, Hollywood and the animation industry, along with Walt Disney's own childhood and vision. The themed sections of Disneyland best illustrated this eclectic melange. Futurism, technology, transplanted European fairytales, American frontier myths, colonial-style adventuring and 1950s suburbia all sat within park berms. Nature remained a key element of the Disney world, but not the only factor. Some of the most famous Disney products, such as *Mary Poppins* (1964) and Volkswagen Beetle-based *The Love Bug* (1968), pushed people-based stories. The majority of characters and stories, however, contained a striking organic contingent. As scholar David Whitley related, 'the theme of wild nature forms the very heartland of Disney's animated features'. It remained at the core of the classic Disney vision.[8]

Disnature

At a simple level, Disnature replicated American nature. In Disney parks, it amounted to a cartoonised collection of monumental scenery (geysers etc.), representative landscapes (prairie, mountains) and signature animal characters. Something that conformed to the expectations of the majority, Disney presented 'mainstream nature'. Disnature also exhibited distinctive Disney properties:

play, adventure, entertainment, harmlessness, innocence, human control, anthropocentricism, overt characterisation, fantasy, reductionism and childishness. Often sold as a brand and a product, Disnature existed within a panoply of toys and memorabilia (read Disneyana).[9]

Disnature drew on Walt Disney's personal experiences of the great outdoors, in particular, growing up as a child on a farm in Marceline, Kansas. Only amounting to a few years, time at Marceline proved to be dramatic in shaping Walt's mindset and direction, and stimulated an enduring fondness for farmyard creatures. Disney related 'When I was a kid, we lived on a farm and we had all kinds of animals. I'd like to have that again.' His daughter, Diane Disney Miller, noted that her father remembered, 'many of the animals on Grandfather's farm better than many of the people he's met. The special feeling he has for animals started then.' Disney went on to surround himself with a cartoon farm on the animated page.[10]

The inner cogs of the Disney corporate machine also fashioned Disnature. Manufactured into merchandise lines, the organic emerged as the ultimate Disneyfied consumer product. Nature became, in essence, a consumer good, something to buy (into). Like nature stores in shopping malls, Disney parks sold fake mementos of the great outdoors. Nature was no longer a resource found on logging and mining frontiers but a bit of plastic sold in a store.

At a cultural level, like everything else, nature was caught in the pull of Mickey Mouse, and totally Disneyfied. Animals were mimicked, sanitised, reduced (often in scale and meaning), and cartoonised. The Disney project remade things: remaking not just the past (rendered 'distory') but also the natural world ('disnature'). The Disney theme park, along with the Disney town of Celebration, Florida, operated as sanitised, safe, and denatured environments. Any wild nature trapped within these geographic locations, or caught on screen, was party to the Disney brush. This Disneyfication of nature transformed terra firma into softly drawn pastel forms. A Disney look to nature emerged, with a specifically cute and appealing Mickey Mouse glow. The natural world was rendered in Walt's/Mickey's own image.

Like any other element within the Disney world, perfectly maintaining 'Disnature' entailed significant control. In cartoons, conformity seemed eminently plausible. As Bob Thomas told, 'cartoons are the most controlled of motion-picture mediums. The animator draws his own characters, makes them move, places them against a background that is complementary. Total control.' In merchandising, Disney maintained tight control over its product range and licensing agreements. Challenges emerged, however, when promoting Disnature within the park system. Bryman related how 'the company revels in its mastery over the natural order and celebrates such things as importing flora from all over the world to create the right kind of vista for its attractions'. At the same time, 'Disnature' appeared prone to irregularities. Environmental factors interfered, and flora and fauna co-opted into the Disney machine went their own way. Plants withered and animals failed to entertain. Total control proved a seductive mirage.[11]

The process of Disnature entailed the filtering, qualifying and determining of modern expectations of nature. The Disney corporation created Disnature as a sanitised product, marking the end of ecological complexity, and the sense of the wild as something beyond us. Despite the prima moda of Disney to tell stories and furnish myths, nature lost its mystery through such an enterprise. The organic became reductive cartoon fun. Disney drew on the psychological power of experiencing things for the first time, its parks and films encouraging a 'childlike' encounter with nature, seeing it as innocent and Disney-like. Within this broader project, nature became wholesome, generic, and reductive. Even the villains seemed harmless. Rather than red in tooth and claw, Disnature was all big ears and fake smiles.[12]

The veritable success of Disnature threatened the demise of real, knowable, material nature. The spread of Disnature seemed akin to an unstoppable exotic, a weed supplanting the native plant. For those disconnected from material nature, Disnature provided a welcome didactic function. It provided a key reference point or educator over what nature meant. With little contact with the outside world but plenty of contact with the world of Disney, 'fake' nature prospered in American culture. One nature replaced the other. Disney recoded popular understanding of the great outdoors.

The triumph of Disnature (or the cartoon fake) left nature in crisis. French philosopher Jean Baudrillard questioned the existence of the original or authentic in the American landscape. He saw America as a symbolic realm with few historic, cultured or authentic markers. Within such a realm, Walt Disney held prime position as a navigation device for a broader American cultural milieu, 'Disneyland is presented as imaginary in order to make us believe that the rest is real, when in fact all of Los Angeles and the America surrounding it are no longer real, but of the order of the hyperreal and of simulation'. A sort of unintended 'architecture of reassurance', to borrow Karal Ann Marling's phrase, Disney (and Disnature) provide saccharine comfort in an unsettled postmodern and symbolic world. Replacing the old world and old nature, Mickey Mouse, Bambi and company found a captive audience.[13]

Mickey Mouse and nature mascots

Mickey Mouse was not the first Disney staple character. The fledgling studio produced numerous cartoons in the 1920s involving Oswald Lucky Rabbit. In 1928, however, Walt Disney visited New York for a distributor meeting that lost the rabbit in a legal wrangle with Universal. According to fable, on a train back to Los Angeles, Walt conjured up a replacement, Mortimer the field mouse (quickly relabelled Mickey). Chief animator Ub Iwerks drew the first Mickey. Sharing some similarities with his contemporary Felix the Cat, Mickey of the late 1920s was, according to Iwerks, 'the standardized thing . . . Pear shaped body, ball on top, couple of thin legs'. Simple to draw, production came more easily. He also resembled early Disney work. Watts commented that, 'Mickey was little more than a reworked version of Oswald the Lucky Rabbit – the ears were shortened, the figure was made more rounded, and the character became more loveable'. November 1928 saw the release of the first official Mickey movie, *Steamboat Willie.*[14]

The Mickey of the *Steamboat Willie* era sported a slapstick style of humour and a cheeky attitude. Playing the archetype of trickster, a cartoon version of Charlie Chaplin, the mouse exhibited an adult-style persona. According to Steven Watts, 'Mickey's antics embodied a carnival spirit of fun that tapped elements from popular music, vaudeville comedy, and dance, and playful social satire to feed the fantastic images that moved magically across the screen.' The mouse tapped mainstream interests and popular pursuits, representing a cartoon synthesis of mass culture. Within just a few years, Mickey emerged as a cartoon 'star'. One Hollywood commentator wrote in 1931: 'Mickey Mouse has a bigger screen following than nine tenths of the stars in Hollywood'.[15]

By the mid-1930s, a degree of tricksterism had given way to

Figure 7.1 *Fantasia* (Walt Disney, 1940).

maturity, as evidenced in *Band Concert* (1934). At the peak of the depression, Disney's mascot provided welcome escapism. He also kept companies other than Disney afloat. A deal with Ingersoll Corporation in 1933 for a Mickey watch saved that company from bankruptcy. Testament to the global appeal of Disney, and the welcoming of Americana, the Japanese emperor wore such a timepiece. Similarly, a Mickey Mouse wind-up train aided Lionel stave off collapse.

In the 1950s, Mickey emerged as the middle-class American of suburbia, the corporate symbol of Disney, a sensible company icon, or, to borrow sociologist William H. Whyte's moniker of the period, 'the organization man'. Artistry of Mickey gradually shifted to match public taste. Mickey became increasingly childlike (or neotenised), with his form and posture relaxing. As Stephen Jay Gould noted, Mickey 'evolved' from a cheeky, almost nasty figure to a lovable, harmless mouse. His cartoon outline shifted from harsh to soft lines. The shifting form of Mickey Mouse proved emblematic of the evolving Disney look and message. Early attempts at original (even dark) works such as the *Skeleton Dance*

(1929) or even the psychedelically informed *Fantasia* (1940) gave way to simpler happy endings and morality tales. In effect, 'cute nature' prospered over more realistic and darker interpretations.[16]

Interpretations of Mickey generally focused on his shifting form and evolution as a cartoon character. Scholar Robert Brockway situated Mickey as a modern-day enigma, claiming, 'He is as complex as Disney was himself and as profound in his symbolic and mythic implications as any mythic or fairy tale character'. Brockway overstated his case but the interface between nature and mouse proved complicated. In technical terms, early Disney animation resembled early animal photography/film whereby a series of sequences or sketches were caught on camera. Both photography and animation, in their primitive stages, highlighted the capturing of movement, as well as pushing the novelty of their visual format. Walt Disney looked to Edward Muybridge for his work on animal and human locomotion.[17]

The original form of Mickey bore some resemblance to a rodent, or at least, to a caricature of a rodent. He was a direct descendant of the farmyard mice that Walt Disney related to at Marceline. Daughter Diane Disney Miller described how Disney had a 'special feeling for mice'. Despite his softening form, Mickey never lost his status as a mouse, an ambassador for a species traditionally treated with fear and derision. Mickey embodied 'approachable' nature: domestic, familiar and innocent. In that regard, he was part of a broader coterie of animal characters in the American market, most notably Snoopy, the Muppets and Sesame Street. Elizabeth Lawrence made the point that 'Entering the human world, pigs, bears, and frogs become approachable and lovable as they become detached from the animal realm through fantasy that generally casts them as juveniles'. Nature was redrawn to fit human expectations and norms and, when rendered youthful, appeared particularly appealing and harmless. This neoteny of Mickey (and other animals) cast the human as dominant and nurturing, and nature as inferior and childlike.[18]

Mickey turned eighty years old in 2008 but displayed no signs of ageing. In a society increasingly drawn to plastic surgery, health cures and the elixir of youth, Mickey Mouse effortlessly maintained his youthful countenance. Walt Disney's own idealisation of childhood in Marceline played out in Mickey 'being the mouse' but also 'being the child' forever. Disney always intended such a juncture. Artists were instructed to think of a boy not a mouse when

drawing Mickey, to 'escort back to a state of youth' those watching the cartoon. Mickey's squeaky voice managed both juvenile
and rodent qualities. The mouse thus always appeared more than
a mouse, a magical rodent intermediary capable of rolling back
the years. Granting the opportunity to see the world (and nature)
for the first time, he embodied a fleeting return to childhood. First
contact with the organic was tied to innocence, purity and revelation. He touched on the social fascination with youthfulness and
discovery. Mickey forever occupied this liminal space. As Lawrence
remarked, 'As one of the most highly anthropomorphized of all
animal figures encountered in present day life, it is not surprising
that virtually no one perceives of Mickey primarily as a mouse, but
rather he is seen as a special kind of youngster.'[19]

Was there ever much mouse in Mickey? The increasingly childlike and cuddly form of Mickey over the course of the last century
might be interpreted as a 'denaturing' of the rodent. More accurately, nature was humanised and recodified by Disney imagineers.
Mickey Mouse, as one of the earliest forms of Disnature, provided
a valuable segue between nature and society, the organic world
and modern media. His rise in the 1930s partly reflected a human
need for nature in industrial and urban society. To some extent,
Mickey served as a cartoon transmogrification, a twentieth-century
American twin to Native American 'creatures' that betrayed both
human and animal personas. He was both mouse and man. As illustrated in *The Disney Magazine* of the 1980s, Mickey even owned
his own fully furnished house and pet dog, and played detective.[20]

Mickey Mouse serves as the premier corporate symbol for the
Walt Disney Company. The use of nature for mascot is hardly a
Disney invention: bears have occupied royal European coats of
arms for centuries. In the United States, the turkey was first suggested as the animal mascot for the nation. What marked Mickey
as exceptional was his global recognition and his corporate
might. Throughout the twentieth century, Mickey operated as a
pre-eminent symbol of American capitalism. The iconography of
the mouse proved diverse and omnipresent. Within Disney park
berms, the staple black ears – unique to Mickey Mouse, no other
rodent – occupied all kinds of spaces. They adorned gifts (such as
the classic Disney-ear headwear) and poster boards. As Lawrence
remarked, you can find 'his form cut into a topiary display, his
face emblazoned on bright colored flags and pennants, his grin
blooming from a flower bed'. Hence, other forms of nature (and

other artefacts) were moulded into Mickey-styled Disnature. This Mickey Mousification of the landscape cemented Disnature as primarily about whimsy (a phrase pushed by Alexander Wilson) and focused purely on play. Disnature was also 'nature' as corporate brand, the theming of an American company around 'the natural world' (albeit in highly caricatured form). The lovable rodent denoted, and disguised, the ruthlessness of American capitalism.[21]

Bambi and the death of (Mother) Nature

Nature always featured heavily in the Disney oeuvre, inspiring noteworthy characters, such as Donald Duck, and early Technicolor cartoons, such as *Flowers and Trees* (1932) that related a love story between two trees, their romance threatened by an 'evil' tree stump and a forest fire. In 1937, Walt Disney gained the rights to Felix Salten's *Bambi* (1923). Salten (also known as Selig Salzmann), a resident of Budapest and Vienna, scribed a right-of-passage story about an orphaned deer fawn with regal ties. The animal story boasted a folkloric feel and put forward an early environmental agenda by its criticism of hunting and exploration of animal fear of humans (the latter termed 'He' in the novel). A translated version appeared in English markets in 1928.

The Disney *Bambi* (1942) deviated in a number of ways from the novel. Scene writers replaced Salten's sombre and serious text with a narrative based on nature's life cycles and seasons. Such an approach to film-making set the standard for later wildlife documentary-makers. The emphasis on life cycles gave Disney the opportunity to highlight rebirth in the natural world, and offer a positive ending to the picture, thus rejecting death as the central aspect in Salten's original. The most radical twist in *Bambi* came in the Disney approach to animation for the feature. *Bambi* pioneered nature realism.

Striking a very different chord to Mickey Mouse, Disnature in *Bambi* was all about realism. Disney demanded the 'capturing' of real animal behaviour on film. Thus, wildlife literally entered the studio. Live fawns (along with rabbits, skunks and owls) provided models for animators. The great outdoors came indoors. Wildlife artist Rico LeBrun taught classes to the Disney team. Live footage of animals served as inspiration. Walt Disney explained, 'One thing always leads to another around here . . . In Bambi we had to get closer to nature. So we trained our artists in animal locomotion and anatomy.' The making of *Bambi* resembled the tireless manufacture of a wildlife documentary. Artist Robert Feild described how

> It was decided that the animals must be allowed to tell their own story as far as it was possible! The continuity must be based not upon what man has read into nature but upon what man can learn from the nature of those things they share in common.

Such an approach forwarded *Bambi* as a proto-ecocentric text. The absence of any notable human characters in the film underlined the focus on wilderness, Bambi very much the 'movie star'. Feild went as far as to claim, 'The whole Bambi unit surrendered to the spirit of the wild'.[22]

There were clearly limits, however, as to how much animal agency could influence a Disney animated feature. Environmental determinism was restricted to the accurate depiction of animal form, not to story content. *Bambi* was Disnature after all. As King commented, 'This is nature, but a very special kind: not an eco-system, but an ego-system-one viewed through a self-referential human lens: anthropomorphized, sentimentalized, and moralized.' Interpreting the filmic content as more about social hierarchy and gender than about nature and the environment, David Payne answered an indefatigable 'no' to 'Is this nature's story?' and felt uncomfortable about how the movie put forward the concept of deer being above predators in the ecological pyramid. Payne also pondered the celebration of patriarchy in the narrative, with the male stag situated as hero, the mother as sacrifice. At other levels, the film served as a story of innocence lost, of a child (Bambi) growing up facing its anxieties over parental separation.[23]

Bambi proved to be a financial disappointment. Released in August 1942, in the midst of war, and with a limited overseas market, the film generated only meagre box-office returns. Costing $2 million to produce, the film garnered $1,230,000. Disney's next big release came some eight years later with *Cinderella* (1950). Two issues dominated contemporary and later coverage: the issue of realism within the film, and Disney's stance on hunting.

The response to Disnature/nature realism in the film varied. *Variety* loved 'the transcription of nature in its moments of turbulence and peace', enthusing how 'Bambi is gem-like in its reflection of the color and movement of sylvan plant and animal life'. Some reviewers pondered the need for a high degree of authenticity: why not just use live action in nature stories? Academic commentators also explored the interface between wildlife and image. For Christopher Finch, the realism of the aesthetic clashed with fantastical situations whereby owls and rabbits, natural enemies, befriended each other in the film. The lack of predators (other than man) in the movie hardly denoted ecological realism. For Payne, however, the wonder of Disney revolved around putting nature on the big screen: 'In the Disney guild, technical and metaphysical ambitions combined to create a vision of the film that allowed the scenic properties of nature to nearly overtake all other elements of the story'. The striking look of Disnature clearly sold the movie.[24]

Bambi provoked controversy by its presentation of hunting. His mother killed by a hunter, Bambi is left heart-stricken by bereavement in a pivotal scene of the film. The latter third of the film is devoted to a chase sequence whereby Bambi takes flight from rabid hunters. Happening offscreen, the death of the mother deer had considerable impact on viewers. Disney succeeded in generating a truly powerful cinematic moment. *Variety* referred to it as 'a scene of deep pathos'. Commonly cited as a vivid childhood memory of the baby-boomer generation, many parents of the day felt the scene too disturbing for their children. The 'death scene' also offended the American hunting lobby. The editor of *Outdoor Life* magazine declared the Disney movie, 'the worst insult ever offered in any form to American sportsmen and conservationists'. By contrast, the Audubon Society compared *Bambi* to *Uncle Tom's Cabin* (1852) in terms of it popularising an issue and serving as a statement of protest. Hunt opponents carried the moniker of Bambi-lovers, hunters Bambi-killers/haters. Ralph Lutts called the 'sentimental, sympathetic attitudes toward wildlife, especially deer', generated by the movie 'the Bambi Syndrome'. Lutts went as far as to declare *Bambi*, 'the single most successful and enduring statement in American popular culture against hunting'. The routine re-release of Disney's *Bambi* certainly made sure new viewers were exposed to the anti-hunting narrative. Considered a piece of animal rights propaganda, *Bambi* broadcast a surprisingly critical voice packaged in Disney sentimentality. According to the film narrative, the American forest was no longer the hunter's paradise, it was more paradise destroyed, marked by the sound of gunfire. The gradual decline in hunting in the twentieth century, however, owed more to changing attitudes toward recreation and the outdoors than to the impact of one film. *Bambi* became a figurehead, even a scapegoat, for broader social and economic forces at work, a trophy animal if you like.[25]

In retrospect, the true significance of *Bambi* was an end of nature text, a breakthrough environmental movie that questioned an American hunting tradition. Disney commented on the relationship between man and nature in a critical but eminently accessible fashion. Given the Holocaust and mass killings of global conflict, misanthropic tendencies were probably reflective of 1940s sensibilities. Man the destroyer, however, crossed clearly into nature's domain via the cartoon forest. *Bambi* served as a story of nature's territories 'ravaged and violated by "Man"'. Disney showed Bambi (and with it, the wider natural world) as innately vulnerable and innocent while humankind appeared all-powerful and destructive. As Whitley noted, 'the idyllic realm of nature rendered vulnerable by human action' proved a key aspect of *Bambi*. Despite (or perhaps because of) the simple, easy to follow line of Disney narrative, audiences related to such a fundamental concept. In particular, the death of the mother deer generated a deep emotional empathy and bond with the fragility of nature. Disney, meanwhile, pushed the notion of nature's balance: that left alone, ecology sought a perfect equilibrium. Man (the hunter) posed a considerable danger to this state of calm. Humanity ultimately threatened the loss of 'ideal' nature. And ideal nature equated to Disnature: pretty, innocent, pastoral and pastel, and Bambi-like. Although *Bambi* ended with an element of optimism and rebirth, Disney provided no answers as to what humankind should do to save the Earth. But it did raise doubts over our role as stewards.[26]

True-Life and the suicide of lemmings

Along with promoting a shift in attitude towards hunting, *Bambi* also spurred new directions within the Disney corporation. Following on from 'nature realism', the studio began to put more 'live action' on the big screen. A range of True-Life wildlife films in the 1950s corresponded with the first Disney 'live action' adventure movies, including *Treasure Island* (1950) and *20,000 Leagues Under the Sea* (1954). The nature series established Disney as a new force in the education market. Daughter Disney Miller imparted that her father 'had always had an urge to do something he thought of as "educational"'. He also recognised the commercial opportunities in wildlife-based documentaries. The True-Life Adventures series began with *Seal Island* (1948), a look at marine mammals off the coast of Alaska. After two years of filming by Alfred and Elma Milotte, their thousands of reels of film translated into a twenty-seven-minute movie. Short in length and with no obvious stars, Disney's distributor RKO initially rejected the picture as an unlikely piece of cinematic revenue. Disney responded by showing the documentary at the Crown Theatre in Pasadena for two weeks. He entered the film for the annual Academy awards. *Seal Island* went on to national distribution, critical acclaim (it garnered an Oscar for Best Short), and financial success.[27]

The True-Life Adventures series proved to be a phenomenal success for Disney Studios. For Derek Bouse, True-Life 'borrowed from other genres – documentary, travelogue, melodrama, musical, even cartoons – in addition to pulling together its own already familiar elements' to generate a successful package. While sporting a mainstream Hollywood feel, Disney added its own signature style to the series. With production costs amounting to the hiring of nature photographers and editing of film, wildlife documentaries proved significantly cheaper to produce than animated features. Naturalists submitted films for consideration, *The Living Desert* (1953) emerging from a ten-minute film by a UCLA (University of California, Los Angeles) doctoral student. A cost-effective format in the uncertain post-war market, the True-Life Adventures series became an unexpected money-spinner for the Disney Studios. *The Living Desert* cost $400,000 to make but income exceeded $4 million. Disney coupled the shorter films with Disney features, *The Olympic Elk* (1952), for example, showing with the re-release of *Snow White*. Both in terms of content and accessibility, Disney nature films generated mass appeal. King goes as far as to claim, 'Disney films, which reached millions of children and their parents in theaters and on television, as well as in the classroom, exerted a cultural influence far wider than Rachel Carson's *Silent Spring* or the Sierra Club'.[28]

Disnature thus reached a significant audience in the 1950s. For a nation keenly embracing cinema and television, the series of Disney pictures added a new genre to the schedule, natural wonders sitting alongside quiz shows, Hitchcock dramas and television westerns. Gregg Mitman referred to the True-Life Adventures series as betraying 'populist leanings', and that 'Disney's nature – benevolent and pure – captured the emotional beauty of nature's grand design, eased the memories of the death and destruction of the previous decade, and affirmed the importance of America as one nation under God'. They were very much 'feel-good' nature films. According to Diane Disney Miller, Disney chose to 'train his field photographers, who were primarily naturalists, in the fine art of pleasing audiences'. Certainly, the Disney True-Life Adventures suited the grey-flannel safety of the period. The True-Life series celebrated home soil and American nature. The documentaries captured the wonder of America on film. Disney also tapped the recreational impulse of the 1950s. Suburbanites hit the open road for weekend wilderness adventures while Disney provided valuable celluloid excursions mid-week. The documentaries boasted middle-class kerb appeal. Reflective of the suburban zeitgeist of the period, *Nature's Half Acre* (1951) highlighted 'back yard' domestic nature. It employed time-lapse photography to show flowers blossoming and uncovered the private life of spiders and honeybees. While no humans featured in the pictures, the True-Life series was very much framed by social mores of the time. As Cynthia Chris noted, 'innovative, ambitious, risky, and influential; they were also sen-timental, anthropomorphizing, and steeped in postwar ideologies of progress and individualism, homeland prosperity, and so-called family values'. Chris labelled them 'animal-populated allegories of proper parenting', with the animal guiding the American parent on how to raise the perfect nuclear family in the 1950s.[29]

The True-Life series fused entertainment with education. Narrators gushed with all manner of wildlife trivia as the action played out in each film. Walt Disney saw the merit in making information accessible, relating that 'We'll make educational films, but they'll be sugar-coated education.' Documentaries entered the American school system, with a whole generation of school kids learning about wildlife courtesy of Walt Disney. The films provided a didactic role and helped children transition from cartoon to real-life animals. Derek Bouse mused, 'It is possible that for a time Disney's wildlife films reached and perhaps influenced more viewers globally than any other nature-oriented media.'[30]

Within the film industry, the True-Life series revised the genre of nature documentary. One specific innovation included the elimination of a human actor on screen. The direction of the filmic narrative followed either an individual animal (or animals) or focused on one geographical region. Prior to Disney, documentaries typically featured a colonial gaze directing the exploration of distant exotic realms. Disney redirected attention to home soil and celebrated American nature. Presentation of the natural world shifted from nature 'red in tooth and claw' to something far more cute and friendly, far more Disney. From sports hunter safari-man Frank Buck's *Bring 'Em Back Alive* (1932) to *Nature's Half Acre*, wildlife on screen morphed from the ultimate peril to domestic ally. Animals became whimsical characters. The True-Life series set the tone for future nature documentaries. For Chris, Disney 'repopularized the wildlife film', and placed nature firmly on to the big screen.[31]

But was it nature or Disnature? And were the films about entertainment or education? Mitman noted that, 'Disney took great pains to emphasize that in these films nature wrote the screenplay; the eloquence, the emotion, the drama were nature's own.' Rather than Mickey Mouse, real nature attracted viewers. But the True-Life series remained very much a Disney product and sported a Disney view of animals. Staff from both *Fantasia* and *Snow White* worked on the live-action productions. The Disneyfication of nature within the documentary ranged from adding music and narration to granting animals their own personalities. Disney transformed raw nature into facile comedy. The 'mysteries' of the sand uncovered in *The Living Desert* included square-dancing scorpions and gender-aware spiders, 'Mrs Tarantula' going by the epithet of the 'lethal lady'. Disney's own comments on an early reel read: 'In sequence where tortoises are courting . . . They look like knights in armor, old knights in battle. Give the audience a music cue, a tongue-in-cheek fanfare. The winner will claim his lady fair.' While authors such as Ernest Thompson Seton granted animals endearing characters, Disney took the 'animal personality' to a new stage. Viewers saw wildlife in real habitats but through a cartoon lens. For Janet Wasko, 'The ultimate value of animals, according to the Disney perspective, is similar to the attitudes some people have towards pets, which are thought of as for personal gratification or for basically entertainment purposes.' The True-Life pictures presented Disnature, rather than real nature, on the screen. In the process, the organic was cheapened, demystified, and even falsified.[32]

In 1958 Disney released *White Wilderness*, a live action film documenting wildlife in the Arctic. In one memorable scene, Winston Hibler narrated the 'lemming legend'. Hibler related 'the legend of mass suicide' surrounding the affable rodent, one of 'nature's mysteries' about to be explored. 'A final rendezvous with destiny and death', lemmings were then observed jumping off a steep cliff into the ocean below, making a desperate swim out into the Arctic Ocean before freezing to death. Caught on film, the avalanche of soil, rocks, and lemmings, for Hibler a 'frenzy' of activity, proved both enthralling and bizarre in equal measure. Diving off the 'final precipice', with their 'last chance to turn back' gone, tiny creatures skydived to their death. Disney cameras provided spectacular footage of the bold jumps. Orchestral music accompanied the scene. This was nature at its most dramatic, revealing and shocking. The lemming suicide transfixed its viewers. The film won the Academy award for Best Documentary Feature that year.[33]

White Wilderness propagated the myth of lemming suicide. The film popularised the idea of animal self-sacrifice. But the scene itself was staged. The film crew pushed lemmings off a crevasse on the Bow River near Calgary, some distance from the ocean. Cameramen were instructed to 'throw them off the cliff by the bucketful'. Allegedly, a rudimentary makeshift machine kept the lemmings from crawling back. It was anything but true life.

As a reliable and trusted format, viewers had no reason to question the presentation of lemmings in the wildlife documentary. They assumed such programming to be accurate. Few had ever seen lemmings in the wild and naturally went by the Disney presentation. Disney shaped popular attitudes towards the rodent. Wildlife documentary-makers always manipulated raw content by cutting scenes and speeding up dramas of birth and death, predation and procreation. As Bouse informed, 'Film and television are about movement, action, and dynamism; nature is generally not.' Such edited snapshots of animal life enabled only a limited view of ecological reality. Televisual expectations shaped the film narrative as much as, if not more than, the natural content. A 'soap opera' or 'action film' of animal characters was the result, with nature cast as media entertainment.[34]

But, for some critics, Disney took the need to entertain too far. Lemmings did not commit suicide. For Mitman, the filmic construction of lemming activity represented the worst of wildlife staging for the camera, and the exploitation of trust in the wildlife documentary. Across the True-Life Adventures series, the addition of music in time with the action, the presence of humorous sketches and commentary, and the formation of 'good' and 'bad' characters challenged the whole concept of a factual documentary. In response to *The Living Desert*, *Variety* noted that the squirrels were 'exhibiting all the charm of a Disney cartoon character', as if Mickey Mouse might appear at any point. A reviewer for *The New York Times* lambasted the documentary as the 'synthetic reconstruction of nature . . . passed off as real'. Most reviewers deemed *The Living Desert* a flawed production. *Time* magazine lamented:

> Thus far, Disney seems afraid to trust the strength of his material. He primps it with cute comment and dabs at it with flashy cosmetical touches of music. But no matter how hard he tries he can't make mother nature look like what he thinks the public wants: a Hollywood glamor girl.

Writing in the late 1960s, Richard Schickel felt that Disney reduced the meaning of nature in the True-Life series,

> wild things and wild behaviour were often made comprehensible by converting them into cutenesses, mystery was explained with a joke, and terror was resolved by a musical cue or a discreet averting of the camera's eye from the natural processes.

Disnature changed nature, made it dishonest. When nature failed to fit the Disney vision, the organic had to change. The core concept of the documentary conflicted with Disney's prime operandi of commercial entertainment. Animal personalities (Disney stars) clashed with the need in wildlife biology to disassociate with the subject matter, to depersonalise the animal. Many True-Life creature entertainers assumed human traits on the big screen. Similar to their cartoon brethren, Disney live animals impersonated the human.[35]

Park-life and peaceable animal kingdoms

Not restricted to celluloid, Disnature also found physical form in the 1950s within the park. In the late 1940s, Walt Disney envisaged a public tour of the Burbank Studios. In 1951, a plan for Burbank emerged. Blueprints put forward a conventional park:

green, rural and European in style, complete with a picnic area, boathouse, lake, bird island, and fairground. Disney employed a naturalistic look for the proto-Disneyland, with more a modern-day Stourbridge than a west-coast Coney Island in mind. The plans conformed to one of the most important rules of the classic park idea: nature as recreation. Subsequent revisions, however, and relocation to Anaheim led to a considerably different outcome.[36]

In July 1955, Disneyland opened to the public. Walt elaborated on his vision for the new park: 'Disneyland will be something of a fair, an exhibition, a playground, a community center, a museum of living facts, and a showplace of beauty and magic.' Notions of wilderness and parkscape had disappeared. With its historic rec-reation of an American Main Street circa 1900 and its 'Wild West themed' frontier land, Disneyland instead paid sentimental homage to a built 'frontier' America. The new park represented 'America only more so'. Disneyland served as geographic celebration of 1950s leisure time, consumption, affluence, and capital. It captured the American Dream and provided its guests with a playful realm of optimism. Tomorrowland postulated a time of artificial living, white cities, nuclear power-fuelled futurism and everything under human control. Fantasyland operated as a cartoon smorgasbord of folklore while Adventureland offered an exotic colonial explora-tion. Disneyana dominated the parkscape. Disneyland provided a wonderfully indulgent giant playscape, one man's fairy tale rendered in plastic and concrete. Inside the park, Americans safely enjoyed the joys of Walt's vision. Outside lurked the Cold War, the nuclear bomb, and pending environmental and social crisis.[37]

Disneyland differed from most parks before it. The exotic, electric-powered opulence of Luna Park at Coney Island was probably closest to the design and look of Disneyland. Created by Frederic Thompson and Elma 'Skip' Dundy in 1903, Luna featured ornamental architecture, bright lights and 'a pageant of happy people'. But the sterility, cartoonery and family focus of Disney made for a wholly different entertainment prospect. Most of all, Disneyland was devastatingly artificial. Bulldozers cleared 160 acres (63 ha) of orange grove for the project. Disneyfication signalled the destruction of nature on a totalitarian scale; nothing was allowed to interfere with the work in progress. In place of rural California, contractors dug artificial lakes and constructed buildings akin to a Hollywood film set. They readied the park for open day. The newly constructed 'ideal America' of Walt Disney featured scant wild flora or fauna – the Wonderful World of Disney had little room for the truly natural.

Historically, the Disney corporation exerted total control over its image, product lines and environment. Inside the park, employers and visitors became automatons. Employees maintained constant Disney smiles and faced instant dismissal for behaviours such as removing their character heads in public. Visitors followed systematic queues and signs. With the potential to deviate from the 'perfect' Disney vision, most flora and fauna failed the Disney audition. Real nature was simply too messy for Disney. Rare bits of flora and fauna allowed inside faced a high level of coercion, party to the Disney look and maintenance regime. Nature was forced to conform to Disnature. Working for Walt Disney, Bill Evans hunted only for 'big' trees that pleased the chief. Disappointed by the flora of Nature's Wonderland Mine Train, Walt commanded, 'Move all those trees back fifty feet. I want the people on the big trains to see what's going on in here.' As a post-war vignette of 'taylorized fun', Disneyland operated as conveyor belt of escapism. Park berms marked a decisive dividing line between Disney artificiality and technology and the wider world. Concrete car parks, a buffer zone of 1950s motor car worship, kept nature out. They also kept Disnature in.[38]

Artificial and reconstructed nature, meanwhile, triumphed inside Disneyland. In the bulldozed soil of orange groves, Disnature bloomed. A self-contained plastic ecosystem provided an artificial nirvana, a habitation zone for all manner of nature-based characters. The environment celebrated the facsimile and simulated. As Karal Ann Marling related, 'in Disneyland, even the natural features of the terrain are unnatural'. Disnature only borrowed from the very best of material nature. Constructed from a steel framework and covered in concrete, the Matterhorn Mountain, added in 1959, mimicked its twin in the Alps. Bryce Canyon national park colours inspired the rock hues for the Disney ride Big Thunder Mountain Railroad (added in 1979). Cartoon characters prowled the park like escaped zoo animals. Disnature provided theatre, spectacle and grand narrative. Disnature was tied to a greater park experience designed to lift mood and generate sales. Senses heightened, visitors partook in a simulated reality trip, relishing an adventure (and landscape) larger than life. It was pure entertainment, Alexander Wilson's 'whimsy' realised. A fantastically fake landscape, Disneyland resembled a larger-scale version of the middle-class suburban lawn, trimmed to perfection and populated by plastic pink flamingos. The adoration of a plastic, techno-logical nature dominated. A Mecca of artificial nature was forged. Disnature prospered inside Disneyland. Outside the park resided the real and inferior organic world (or what was left of it, as new suburbs and freeways took root in outer Los Angeles).[39]

Disneyland was a paean to technology, control and the physi-cal realisation of Disnature. The paean, however, exhibited limits, one example being the Jungle Cruise. Disney's documentary, *The African Lion* (1955), inspired the ride. As Walt Disney explained, 'It all started with an idea that sprang from our True-Life Adventure films. We would duplicate in Disneyland park actual scenes and settings from this nature series.' Such a statement implied a strong wilderness influence on park design. The end product, however, was anything but natural. Nicknamed 'foam-rubbersville' by critics, the Jungle Cruise appeared a very rudi-mentary facsimile of nature. In such rubbery guise as animatronic crocodiles, artificial Disnature hardly seemed better than material nature. Yet the fake seemed enjoyable enough. Fake nature mostly did as good a job, if not better, than its material counterpart. The crocodile opened its mouth on cue. Marling situated the comfort-able design of Disneyland as the 'architecture of reassurance'. Traditional parkscapes had managed such an effect with recourse to a cultural yearning for wild nature. Disney managed this with Disnature, the artificial, and the rubber crocodile.[40]

Opened in 1971, Disneyworld in Florida simply scaled up the Disnature experiment or, in McDonald's slang 'supersized' the endeavour. Employing aliases to keep prices down and avoid land speculation, the Walt Disney Company purchased a total of 27,443 acres (10,800 ha) of land. Floridian swamp then gave way to artificial parkscape. Complete with hotels and shopping malls, a new cartoon metropolis arose from the swampland. Beneath Disneyworld, and hidden from consumer view, a network of underground tunnels (or utilidors) that connected workers and supplies. In 1982, the corporation added Epcot. As a vision of future society, Epcot glorified the technological, and realised the 'white city' model so popular in the post-1945 era. Very little wil-derness featured in such a template of Utopia. Within the broader world of Disney Florida, what nature remained was forced to abide by the Disney vision. With bark from Cyprus trees colouring the water, Disney's Bay Lake lacked the clean sparkle associated with the cartoon look. Corporate employees removed the trees and drained the lake, then imported sand and created beaches.

Added to Disneyworld in 1998, Animal Kingdom served as fitting testament that Disney was close to perfecting Disnature. Disney Chief Executive Michael Eisner delivered the dedication,

> Welcome to a kingdom of animals ... real, ancient and imagined: a kingdom ruled by lions, dinosaurs and dragons; a kingdom of balance, harmony and survival; a kingdom we enter to share in the wonder, gaze at the beauty, thrill at the drama, and learn.

Five hundred acres (200 ha) in area, the park replicated the African savanna and Asian rainforests on American shores. The simulated landscape included rides, conservation stories, safari and shopping opportunities. At the hub of the park resided the Disney Tree (Tree of Life).

The Tree of Life referenced earlier artificial park flora, such as *Disneyodendron eximus* located at the Swiss Family Treehouse in Disneyland. Granted its own scientific name, *Disneyodendron* threatened traditional taxonomy by positing the inclusion of artificial nature in the Linnaean system. In park geography, the Tree of Life served as the naturalistic twin to Sleeping Beauty's Castle found at other Disney sites, the pivotal park 'wienie' or visual treat that every park visitor headed towards. Like a single redwood giant, transplanted to foreign climes, and towering above the landscape, the Tree of Life shouted of the exotic and the monumental sublime. Sculpted on the 'bark', 325 animal shapes from Galapagos tortoises to dolphins. The animal forms celebrated the variety of nature, the diversity of its taxonomy. Deep inside, the tree-trunk theatre showed *It's Tough to Be A Bug!*, inspired by *A Bug's Life* (1998). As a modern-day totem of technology, play and Disnature, the Disney Tree had few rivals. Plastic nature was perfected in the form of 102,583 vinyl leaves. But ultimately, the fourteen-storey naturalistic skyscraper situated nature as a wonder of technology. It provided a beacon of entertainment for humanity, a celebration of simulacrums. Visitors wondered not at the nature on display but at how Disney made nature anew.

The Disney Tree rendered Disnature in gigantic form but still, in essence, it amounted to dead nature, inert and plastic. The variety of merchandise stores littering the Disney jungle sold the same type of nature but in manageable, take-home and pocketworthy forms. Disnature equated to token memento, park commodity. Animal Kingdom represented a Jungle-themed shopping opportunity, a nature-themed capitalist adventure. Instead of the colonial hunter amassing furs and trophies from the hunt, the intrepid Disney traveller took home plastic tigers and dolphin key fobs. Such rampant, Disneyfied commercialism marked the Animal Kingdom as a decidedly opportunistic affair. Disney tied the park experience (and with it Disnature) to the purchase of memento. The practice of hybrid consumption meant little break between shopping. Magical and exotic, the Animal Kingdom was also decidedly ordinary and mall-like.[41]

One key challenge of the Animal Kingdom was successfully to bring live nature into the park. Previously, Disney had selected only the most docile creatures for its parkscapes. Themed centres, such as Seaworld, demonstrated the potential for live displays. At Animal Kingdom, Disney Imagineers assembled a Disneyfied zoo. The corporation collected 200 species from around the world as stock. Like a zoological collector, Disney identified desirable exhibits and endangered species. Transport accidents occurred, including the controversial death of a black rhino. Akin to 'cast members', fauna were enrolled into the entertainment schema, destined to join the park ride.

On Animal Kingdom's celebrated Kilimanjaro Safari Ride, park visitors embarked on a journey to simulated Africa. The ride equated to Indiana Jones meets wildlife documentary. Imagineers furnished an immersive safari landscape akin to a giant interactive diorama. Disney's ride amounted to a modern take on Carl Hagenbeck's animal displays of a hundred years earlier, providing a theatre of nature to watch and enjoy. The safari supplied a tightly orchestrated sojourn through an exotic world. Disnature represented escape and adventure. As Scott Hermanson noted, 'In contemporary culture, nature already exhibits the qualities of a theme park, a fantasyland where we escape the rigors and frustrations of modern life.' This 'managed exoticism', a term of Giroux's, referenced colonial exploration. The Kilimanjaro safari was all about 'visiting' (and conquering) the 'new' world. It represented a fleeting diversion from reality, a speedy jeep ride through a wild and exotic land of ostriches and elephants.[42]

Even on safari, Disnature remained cute and harmless. The

Disney parks promoted nature as intrinsically unthreatening and family friendly. Disnature provided simulated thrill and artificial danger. Hence, rubber crocodiles on the Jungle Cruise and real-life rhinos on the Animal Kingdom were much the same. Both represented simple cogs in nature-themed roller-coaster rides and entertainment machines providing cheap thrills to guests. Citizens entrusted Disney with their safety, confident that the corporation maintained complete control of all factors, including the behaviour of wild animals. Championed and tested at Animal Kingdom, Disney seemed capable of marshalling a total environment, from tree growth to lion denizen. Disney succeeded where others had failed, the human quest for control over nature won out in the modern theme park. In the process, nature was left without agency, transformed into idle plaything and taken for a ride. Wild animals were 'under Disney direction', akin to actors on a stage. As Mitman pithily surmised, 'Here nature is scripted'.[43]

Disney sported a distinctive script for nature. The Animal Kingdom projected a world of synthesis, compatibility and concord. Left to it (or, more accurately, controlled by Disney), the members of the natural world automatically 'got on'. Disnature was assumed safe partly because of this projection. The park appeared a place of ecological harmony and balance. Disney's Animal Kingdom served as a modern, three-dimensional incarnation of Edward Hicks's *The Peaceable Kingdom* (1826). Hicks's painting depicted a wilderness scene of warmth rather than hostility, with Indians and Americans deep in conversation, babies nestled with lions and leopards, and black bears nosing cattle. Disney supplied a similarly comforting visual of peace on Earth, but with a few additional actors. Significantly, not just 'live' animals got on in Disney's Kingdom but all elements of Disnature: wild, domestic, animatronic and plastic. No difference existed between the artificial and the real on the Kilimanjaro safari ride. Disney seamlessly blended fakery with authenticity. Immersed in the entertainment landscape, the tourist failed to navigate the distinct elements. The lack of clear, definable boundaries between the two, the artificial and the real, led to a collapse in differentiation. Disney heralded the death knell for nature as an independent park variable. Something extra was always needed, a human hand or a machine arm, to complete the picture. The interface between man, nature and machine on display at Disney referenced intellectual musings on the fate of nature in the (post)modern world. The cyborgisation of the natural world, written about by Donna Haraway, linked with the audio animatronic world of Walt Disney. Disney championed machine–nature hybridisation under the twin banners of entertainment and consumption. Mickey Mouse pointed with his white-gloved hand to future, or post-nature as, at the very least, hybridisation, or worse, cartoon mutation.[44]

Dis-connect and the death of the mouse

The death of nature in the United States is marked by stages. The shift from rural to city spaces, the demise of working with the land, and the rise of an industrial landscape embodied the first key shift in American life. Wilderness disappeared and animal extinctions followed. Workers struggled with the loss of a daily, immersive and tangible relationship with the natural world. Modernity spelt the swansong for intimate human–nature relations. This sudden lack of contact with real nature in the late nineteenth century marked the first stage of the death of nature.

Through the twentieth century, Americans increasingly relied on media, rather than outdoors experience for their connection with the natural world. Nature was reduced in the process, turned into 'image': first paintings, then film, and then computer-generated facsimile. Jonathan Burt saw the 'image' stage as 'palliative, consoling us for a sense of loss and rupture'. A celluloid/canvas comfort blanket filled the vacuum left by nature's disappearance. The demise of 'real' nature gave way to the rise of 'image' nature. An intimate relationship with the image replaced an intimate relationship with the real. Stage two had begun.[45]

Disney perfected 'image' nature. The corporation aided the shift in attention from real to fake, helped distract from the death of material nature, and rallied the victory of simulacrums. King related how

> Our reliance for nature's image and context shifted from first-hand experience to the novel, western-school painting, and nature photography, culminating with the film and television versions that were shaped, and continue to be influenced by, the Walt Disney Company's animated films and its live action True-Life Adventure series of the late 1940s.

Disney produced a mainstream, family-friendly 'image' nature. It provided, 'A near-perfect Technicolour simulation of our media-soaked creation of nature'.[46]

Disnature also codified the narrative of loss. For Wilson, Disney's True-Life Adventures posed as 'documents of a culture trying to come to terms with . . . the end of nature'. Viewing the death of nature, Lippit commented on how 'technology and ultimately cinema came to determine a vast mausoleum for animal being'. Disney contributed to this mausoleum. Like Disneyland's Main Street providing a nostalgic trip to the 'small town America of past', Animal Kingdom did the same for wilderness. Like a realised *Jurassic Park* (1993), the Animal Kingdom provided an escape to a world that no longer existed, brought back by technological magic for the sake of entertainment. The Disney dinosaur ride 'Countdown to Extinction' similarly cemented the death of nature in its ride to extinction.[47]

Offering a multitude of cartoon animals to replace their organic brethren, Disnature attempted to fill the vacuum created by loss. But by advancing image nature, Disney actually furthered the process of dis-connection at work. True-Life wilderness on screen exaggerated the notion of nature as escape, and something unreal and detached. Cartoons lacked any authentic animal voice or agency. Plastic plants supplanted material nature within the Disney park system. Disnature replaced nature, pushed it to one side. With its Tree of Life and cartoon prodigy, Disney popularised the victory of the facsimile.

Giroux described how 'Disney's power lies, in part, in its ability to tap into the lost hopes, abortive dreams, and utopian potential of popular culture'. American national parks once serviced similar needs, providing geographical vignettes of cultural nationalism and hopes of a past age. They functioned as slices of 'primeval wilderness' for the American tourist. In fewer than ten years from opening, total attendance figures for Disneyland surpassed those of Yellowstone National Park, dedicated in 1872. In the 1950s, Disneyland usurped Yellowstone as the nation's pilgrimage site. The cultural meaning of America passed from iconic wilderness experience to plasticity and Disnature. A cartoon mouse led Disney to triumph over nature, with Disney entertainment deemed more important than American wilderness.[48]

But was Disnature a worthy replacement for nature? Aside from the wonder at technology and simulacrums, Disnature was mostly something anodyne, easy on the eye, and a facile creation. On a visit to Disneyland in 1958, Julian Halevy compared Disneyland to Las Vegas. Both 'cheap formulas packaged to sell', Halevy found the blatantly transparent artificiality of Vegas more impressive. When EuroDisney first opened outside Paris, one French critic warned of an imminent 'cultural Chernobyl', Disneyfication of French culture about to spread across the country akin to radioactive fallout. Disnature might be taken as a similar 'Chernobyl'-style accident for the countryside, threatening the collapse of traditional concepts of the organic, Disneyfying (or genetically altering) the meaning of nature and wilderness. Disney spells Dis/dystopia and natural disaster. Americans thwarted plans for a Disney 'America' park in the 1990s because of fears over a Disneyfication of history. Distory was rejected but Disnature has leaked into the popular consciousness.[49]

In the 1980s, in an article entitled 'Disney's Ageless Mouse', Lawrence enthused how Mickey, 'created . . . well over 50 years ago, today seems more alive than ever'. The mouse may not look any older in the twenty-first century but his health is deteriorating. Disnature may now be entering decay. Disney has moved on, embracing adult-geared entertainment shows and movies, as well as capturing the teen market with programming such as *High School Reunion*. The Disney website is little different from ABC or any other television channel. The prominence of cartoon animals has ebbed, even mascot Mickey fading from view. Some might welcome the demise of Disnature, of 'image' nature. The demise of the imagined and constructed marks the end of the second stage of nature. What awaits in the third stage: the rebirth and recovery of 'authentic' nature or a furthering of its demise? The death of Mickey Mouse threatens a period of unbridled uncertainty.[50]

Notes

1. Umberto Eco, *Travels in Hyper-reality* (Basingstoke: Picador, 1986), p. 44.
2. Mark Hamilton Lytle, *The Gentle Subversive: Rachel Carson, Silent Spring and the Rise of the Environmental Movement* (New York: Oxford University Press, 2007), p. 168.
3. Bill McKibben, *The End of Nature* (London: Viking, 1990), p. 4.
4. Garrett de Bell, ed., *The Environmental Handbook* (London: Friends of the Earth, 1970); *Earth* originally the work of Brit Alastair Fothergill for BBC/Greenlight.
5. WD in J. P. Telotte, *Disney TV* (Detroit: Wayne State University Press, 2004), p. vii; Alan Bryman, *The Disneyization of Society* (London: Sage, 2004), p. vii.
6. Finch, in Janet Wasko, *Understanding Disney: The Manufacture of Fantasy*

(Cambridge: Polity, 2001), p. 183; Elizabeth Bell (1995) et al., eds, *From Mouse to Mermaid: The Politics of Film, Gender and Culture* (Bloomington: Indiana University Press, 1995), p. 45; Wasko (2001), pp. 114–16.

7. Henry Giroux, *The Mouse that Roared: Disney and the End of Innocence* (Lanham: Rowman and Littlefield, 2001), pp. 4–5, 89; Bell (1995), pp. 45, 7.

8. David Whitley, *The Idea of Nature in Disney Animation* (Aldershot: Ashgate, 2008), p. 1.

9. On occasion, it drew on nature located outside America, in the form of colonial adventure, as demonstrated in the film *The Jungle Book* (1967), or in generic, unidentifiable realms (*A Bug's Life* 1998).

10. WD in Watts (1997), p. 5; Diane Disney Miller, *Walt Disney: An Intimate Biography by his Daughter* (London: Odhams Press, 1956), p. 16.

11. Bob Thomas, *Walt Disney: an American Original* (New York: Simon and Schuster, 1976), p. 16; Bryman (2004), p. 137; Naturally, the exceptionalism of Disnature might be questioned by situating it within a broader history of the culture of nature. Jonathan Burt noted how 'seeing' is how most of us interpret the animal world. Burt elaborated: 'the position of the animal as a visual object is a key component in the structuring of human responses towards animals generally, particularly emotional responses'. Animals have always proved commonplace in visual culture, from cave paintings to digital pets. With the advent of modern media, especially film, the visual became the key determinant in our view of nature. To a degree, the camera/film became the intermediary between animal and human relations. Disney was thus part of a broader shift in how humanity visualised animals. Disney was also part of the social construction of nature. Humans give meaning to their experiences with flora and fauna, viewing them through a social/cultural lens. Despite their own behaviours, animals are 'given' characters by us. The Disney lens is not that different from our own lens. Jonathan Burt, *Animals in Film* (Reaktion, 2004), p. 11.

12. Giroux (2001), p. 89; Bell (1995), p. 46.

13. Jean Baudrillard, *Simulacra and Simulations* (Ann Arbor: University of Michigan Press, 1994), p. 25; Karal Ann Marling, ed., *Designing Disney's Theme Parks: The Architecture of Reassurance* (Montreal: Canadian Centre for Architecture, 1997).

14. Iwerks in Robert Brockway, 'The Masks of Mickey Mouse: Symbol of a Generation', *Journal of Popular Culture* 22/4 (spring 1989), p. 27; Steven Watts, *The Magic Kingdom: Walt Disney and the American Way of Life* (Boston: Houghton Mifflin, 1997), p. 30. On Mickey, also see Miriam Hansen, 'Of Mice and Ducks', *Southern Atlantic Quarterly* 92/1 (winter 1993), Mary Bancroft, 'Of Mice and Man', *Psychological Perspectives* (autumn 1978). The film premiered with sound, a shrewd late addition thanks to the intervention of Walt Disney. The technical innovation had previously made the film *The Jazz Singer* (1928) a box-office sensation.

15. Watts (1997), p. 33; Louella O. Parsons in Watts (1997), p. 37.

16. Stephen Jay Gould, 'Mickey Mouse Meets Konrad Lorenz', *Natural History* (May 1979). On storytelling, see Russell Merritt, 'Lost on Pleasure Islands: Storytelling in Disney's Silly Symphonies', *Film Quarterly* 59/1 (autumn 2005).

17. Brockway (1989), p. 26; also see Burt (2004), p. 108.

18. Miller (1956), p. 91; Elizabeth Lawrence, 'In the Mick of Time: Reflections on Disney's Ageless Mouse', *Journal of Popular Culture* 20/2 (autumn 1986), p. 67.

19. Lawrence (1986), p. 66.

20. *The Disney Magazine* (London: WD Productions, 1981), British Library.

21. Lawrence (1986), p. 65.

22. WD in Eric Loren Smoodin, ed., *Disney Discourse: Producing the Magic Kingdom* (Routledge, 1994), p. 56; Feild quoted in David Payne, 'Bambi' in Bell (1995), p. 139. Also see Kathy Merlock Jackson, *Walt Disney: a Bio-Bibliography* (Westport: Greenwood, 1993), pp. 41–5.

23. Margaret King, "The Audience in the Wilderness," *Journal of Popular Film & Television* 24/2 (summer 1996), p. 4; Payne (1995), p. 141.

24. *Bambi* review, *Variety*, 31 December 1941; See Christopher Finch, *The Art of Walt Disney* (1973); Payne (1995), p. 138.
25. *Variety*, 31 December 1941; *Outdoor Life*, quoted in A. Waller Hastings, 'Bambi and the Hunting Ethos', *Journal of Popular Film and Television* 24/2 (summer 1996), p. 53; Ralph Lutts, 'The Trouble with Bambi: Walt Disney's Bambi and the American Vision of Nature', *Forest and Conservation History* 36 (October 1992), p. 160; Cartmill felt that *Bambi* amplified existent sympathies for deer, and had a 'profound impact on American attitudes toward hunting, wildlife and nature', Matt Cartmill, 'The Bambi Syndrome', *Natural History* 102/6 (June 1993).
26. Bell (1995), p. 141; Whitley (2008), p. 3. Also see Lutts (1992), David Ingram, *Green Screen: Environmentalism and Hollywood Cinema* (Exeter: University of Exeter Press, 2004), p. 19.
27. Miller (1956), p. 187.
28. King (1996).
29. Gregg Mitman, *Reel Nature: America's Romance with Wildlife on Film* (Cambridge, MA: Harvard University Press, 1999), pp. 118, 110; Miller (1956), p. 189; Cynthia Chris, *Watching Wildlife* (Minneapolis: University of Minnesota Press, 2006), pp. 28, ix.
30. WD in Ralph Lutts, ed., *The Wild Animal Story* (Philadelphia: Temple University Press, 1998), p. 13; Derek Bouse, *Wildlife Films* (Philadelphia: University of Pennsylvania Press, 2000), p. 69. Also see Derek Bouse, "False Intimacy: Close-ups and Viewer Involvement in Wildlife Films," *Visual Studies* 18/2 (2003).
31. Bouse (2000), p. 62; Chris (2006), p. xxi.
32. Mitman (1999), p. 110; Wasko (2001), p. 146; script in Richard Schickel, *The Disney Version: The Life, Times, Art and Commerce of Walt Disney* (Lanham: Ivan Dee, 1997 [1968]), p. 288; Wasko (2001), p. 149.
33. *White Wilderness* (1958) script.
34. Bouse (2000), p. 4.
35. Bouse (2000), pp. 67, 68; Mitman (1999) pp. 120–1; Schickel (1997 [1968]), pp. 288, 52; Mitman (1999), p. 121.
36. For plans, see Marling, (1997), p. 39. Also see Carl Hiaasen, *Team Rodent: How Disney Devours the Wild* (New York: Ballantine, 1998).
37. WD quoted in Michael Sorkin, ed., *Variations on a Theme Park: The New American City and the End of Public Space* (New York: Hill and Wang, 1992), p. 206.
38. WD/Evans: Thomas (1976) p. 264; Giroux (2001), p. 39.
39. Marling (1997), p. 29; Wilson (1991), p. 182. Also see John Hench, *Designing Disney: Imagineering and the Art of the Show* (New York: Disney, 2003), p. 56.
40. WD in Chris (2006), p. 30; Marling (1997), p. 109.
41. See Jennifer Price, *Flight Maps: Adventures with Nature in Modern America* (New York: Basic, 2000).
42. Scott Hermanson, 'Truer than Life: Disney's Animal Kingdom', Mike Budd, ed., *Rethinking Disney: Private Control, Public Dimensions* (Middletown: Wesleyan University Press, 2005), p. 201; Giroux (2001), p. 41.
43. Mitman (1999), p. 1.
44. See Donna Haraway, 'A Cyborg Manifesto: Science, Technology, and Socialist–Feminism in the Late Twentieth Century', in *Simians, Cyborgs and Women: The Reinvention of Nature* (New York: Routledge, 1991), pp. 149–81.
45. Burt (2004), p. 27.
46. King (1996); Hermanson (2005), p. 225.
47. Wilson (1991), p. 155; Lippit p. 187.
48. Giroux (2001), p. 5.
49. Julian Halevy, 'Disneyland and Las Vegas', *The Nation* (7 June 1958); Karen Jones and John Wills, *Invention of the Park: From the Garden of Eden to Disney's Magic Kingdom* (Cambridge: Polity, 2005).
50. Lawrence (1986), p. 65.

The Doomsday Machine

Machine technology

PRESIDENT WILSON. Six months ago, I was made aware of a situation so devastating that, at first, I refused to believe it. However, through the concerted efforts of our brightest scientist, we have confirmed its validity. The world, as we know it, will soon come to an end.

2012

In the film *2012*, science provided a means of monitoring doomsday, and technology a way of avoiding it. An American scientist learnt of the Earth's core overheating while advanced technology, built in China, offered four Noah's Arks to fend off the coming storm. Together, science and technology delivered salvation, circumvented the end of the world. The reality of the disasters studied here paints a different picture. Instead, science and technology helped create the symbolic antithesis of Noah's Ark: a doomsday machine. Along with political misdirection, a capitalist structure, and an immersive catastrophe culture, science and technology served as two driving forces behind disaster, providing the grim mechanics of inviting doomsday.

As we have seen, the advent of new technology regularly serviced narratives of conquest and environmental despoliation in the United States. On the nineteenth-century frontier, the firearm

207

Figure C.1 *2012* (Columbia Pictures, 2009).

Figure C.2 Trinity Test (1945) (courtesy of National Nuclear Security Administration/Nevada Site Office).

(particularly the latest repeating rifles) facilitated the dispatch of millions of bison, passenger pigeons and Native Americans. Other examples include the cotton gin, enabler of mass industrial agriculture and monoculture on the plains, and the combustion engine that kick-started Autogeddon and oil obsessiveness. New chemical agents in the twentieth century, such as DDT, promised Utopia but, instead, delivered poison. From the subtle and invisible, to the blatant and practical, science and technology contributed to a doomsday direction.[1]

Also in the twentieth century, the United States invented the breakthrough doomsday device. In July 1945, Manhattan Project scientists successfully exploded the first atomic bomb, codenamed 'the device', in the Jornada del Muerto desert (translated Journey of Death) near Alamogordo, New Mexico. Author P. D. Smith called the international cadre of scientists gathered on the Manhattan Project 'the doomsday men'. In 1950, on NBC, Manhattan physicist Leo Szilard nicknamed the new H-bomb the 'doomsday machine'. The core machine of Cold War politics and technology, the bomb was designed to intimidate and destroy. In a matter of minutes, it could deliver worldwide environmental doomsday. Americans hoping to live through a nuclear attack were armed with shelters, civil defence drills, a 'doomsday scenario' handbook, and watched Doom Town explode live on television. The atomic bomb reshaped notions of catastrophe and environmental disaster. Its fallout included not just nuclear accidents, such as Three Mile Island, but *Silent Spring*, the 'Population Bomb', and even responses to the Santa Barbara oil spill. Only 9/11 has equalled its influence.[2]

Directing the doomsday machine

With the atomic doomsday machine, issues of control and responsibility became paramount. While fearful of other countries gaining the device, nuclear nations entrusted computer technology with an overseeing task, and considered giving technology the choice of triggering the end of the world. On the United States side in the 1950s, government adviser Herman Khan put forward the need for a computer that controlled the American nuclear stockpile, automatically activating a strike on the Soviets on sensing high radiation in the American continent. He spelled out his ideas for auto-retaliation in the controversial *On Thermonuclear War and Thinking the Unthinkable* (1962). The satirical film *Dr. Strangelove (or how I learned to stop worrying and love the bomb)*(1964) explored the ramifications of this technologically determined self-destruction. 'The whole point of the doomsday machine is lost if you keep it a secret!' cried Dr Strangelove, 'Why didn't you tell the world?' Sociologist Lewis Mumford labelled the bomb 'the central symbol of this scientifically organized nightmare of mass extermination'. Mumford also cast the technocratic nuclear powers of the United

States and the Soviet Union as 'megamachines'. In the 1980s, in response to the open sabre rattling of President Ronald Reagan and talk of an American-based Star Wars defence system, the Soviets developed a secret computer 'doomsday machine'. Perimeter (also known as Mertvaya Ruka, or Dead Hand) served as an Emergency Rocket Communications System from 1985 onwards and provided airborne readiness to strike. On disappearance of (or more accurately, no contact with) the Kremlin or Defence Ministry, Dead Hand enabled an automatic counterattack. If links broke with the Soviet General Staff, and radiation levels rose, Dead Hand launched a range of command missiles that, in turn, initiated the launch of other missiles that brought the Apocalypse. Technophile *Wired* magazine later revelled in the existence of 'an actual doomsday device – a real, functioning version of the ultimate weapon, always presumed to exist only as a fantasy of apocalypse-obsessed science fiction writers and paranoid über-hawks'.[3]

Concern over who or what controlled atomic weaponry reflected broader anxiety over the end of the world. Entrusting computers with overseeing such tasks highlighted the fundamental shift in the modern age from us controlling technology to that of technology controlling us. In a 1967 episode of *Star Trek*, entitled 'The Doomsday Machine', an advanced computer system facilitated destruction of both sides in wartime. In *The Terminator* film series (1984–2009), the machines, Skynet, take over, and unleash a nuclear attack on humanity. The subsequent war against 'the machines' takes on a last-stand dimension, pitting the survival of the organic versus the victory of the technological. Automation, artificial life and dependence on computers triggered fears of technological revenge, a machined end of days. Fears over the end of the world on the eve of a new millennium focused on a mystery 2K bug and computer-triggered self-destruction. From Frankenstein to Terminator, technology has driven disaster and honed fears of doomsday.

In the mainstay, however, Americans themselves have directed the doomsday drama. Religion, politics, capitalism and catastrophe culture guided the process. Manifest destiny facilitated conquest on the frontier, a biblical God guiding the conversion of Native Americans and the wiping out of species on the Great Plains. Scholar Lynn White noted the propensity of the Christian faith continually to service dominionistic and anti-nature thoughts. Broader faith in man (the scientist, the politician, the soldier) to solve any predicament also rendered environmental problems temporary and fixable. Politics shaped the energy landscape, determined national policy, fuelled oil, chemical and nuclear expansion while, in the case of the Reagan and Bush eras, proved resistant to environmental strategies. In the Cold War, bomb development ultimately came from presidential authorisation. Political analysts developed concepts such as Mutually Assured Destruction (MAD) that revealed doomsday-like mentalities in high circles and situated doomsday logic in official doctrine. Doom Town was part the madness. Foreign policy decisions, based around concern over the rise and fall of the United States in global positioning, triggered wars and disasters. Allied to the global outlook and feelings of insecurity, fears of disaster gradually became associated with external threats. In the colonial period, Americans feared the wilderness surrounding them. As in the case of Jamestown, dangers lurked close-by. In the twenty-first century, with fears of nuclear attack, poison gases and global warming, the danger shifted in scale and distance. Anxiety over how far these 'external threats' could be controlled marked the modern era. The disasters in *Inviting Doomsday* suggest a need finally to exact social and rational control over science and technology, not to leave control to politicians, militarists, corporate executives, nor wargame computers.[4]

Fuelling the machine

Unbridled capitalism fuelled the American doomsday machine, contributing to the disasters at Santa Barbara, in the Gulf of Mexico, and more. Expanding markets and mass consumption justified the search for natural resources, such as minerals, oil, uranium, and furs, and the constant need for environmental exploitation. Popular myths of limitless energy and superabundance, and the lure of high profit made sure that the capitalist enterprise continued. And capital depended on a fully functional 'energy landscape' to supply it. The unending quest for the almighty dollar kept the doomsday machine well powered.

The American energy landscape moved through successive stages: first, in the colonial era, cotton and tobacco crops; then, on the frontier, gold and timber materials; then in the twentieth century, nuclear and oil supplies. Initially, images of abundance and profit attracted merchants and settlers to Jamestown and New England but, without the technology or numbers to do significant damage, the results proved limited geographically. By the nineteenth century, Americans proved far more adept at profit making and resource extraction, and took vast supplies of raw material: from minerals to forests to animals. Americans on the frontier anticipated boom periods in the new country. A richness of possibilities had led them there. Meat and timber powered the machine. Profits rose, the nation entered the world stage, but resources quickly became exhausted and species wiped out. The land was no longer a source of easy abundance. In the twentieth century, attention shifted to the potential of new technologies to provide energy and profit: nuclear and chemical. Some, like nuclear energy, failed to provide abundance of money or energy. Others, such as pesticides, made money and furnished crops but exacted significant costs. Motifs of abundance (or, in McDonald's language, supersizing) continued with modern mass consumption. By the twenty-first century, attention again shifted, to expansion into the virtual and technological frontier, with new hopes of limitless profits.

The boom-and-bust system of capital, the rise and fall of markets, devastated whole environments. Frontierism encouraged a conquest narrative exhibiting little concern for the environment. Nineteenth-century mineral rushes demonstrated the environmental process in microcosm: the overnight rise of the mining town, its explosion, and its disappearance. Rhyolite, Nevada boasted a population of nearly 8,000 in 1908. By 1920, the population numbered in single figures, with a denuded, despoiled landscape the result. The boom of capital historically and economically created the conditions for the bust. For the environment, both stages proved deadly, as seen in the development of oil, nuclear or chemical supplies and the disasters that followed. The 'shifts' of capital could also be seen in demography: the leaving behind of the rural landscape and the move to the city at the advent of the twentieth century; the failure of the rural environment to provide (spectacularly witnessed in the 1930s Dust Bowl); to, more recently, the death knell of the American metropolis, as witnessed at New Orleans.

Capitalist ways of organising production also regimented life as well as reshaping the environment. In the early twentieth century, Fordism encouraged a conveyor-belt system of workers singularly dedicated to one task. In the late twentieth century, McDonaldisation fine-tuned the process, co-opting customers into servitude. At first, African slaves and cotton crops made up the doomsday; now we are all slaves to the doomsday machine.

Catastrophe culture

The doomsday machine thrived in a specifically American culture of catastrophe. Fear and anxiety resided in the construction (and belief in) imaginative realms of national failure and destruction. A culture of catastrophe first gained agency in the 'starving time' at Jamestown. The trauma residue lingered, subsequent disasters fashioning a collective sense of vulnerability that a new nation could be torn down as easily as built up. Gradually Americans came to exhibit a specific psycho-geography, an attitude to disaster that feared collapse but also invited it. Many pushed to one side the environmental risk of new frontierism and development. A small minority joined catastrophe cults, doomsday groups spouting millennialism.[5]

As witnessed in *Inviting Doomsday*, catastrophe culture emerged from a range of factors. In basic form, it fused anxiety about nature, God, humanity and later technology. Responding to unexplained events, such as comets, as well as the impact of Euro-Americans, Indian nations first predicted the end. Stories of great floods proved common in early world culture. Euro-American fear of environmental catastrophe and doomsday events heralded from the Book of Revelation's End of Days, and religion continued to promote an 'end of days' prophecy in the form of The Rapture up to and past the new millennium. At Jamestown in the 1600s the colonial experiment exhibited catastrophe in imagination as well as practice. Caught up in folklore, superstition, and the dangers of the New World, New Englanders saw evil in the wilderness and feared being overwhelmed by savagery, plague or natural disaster. In nineteenth-century America, fear shifted, with the effects of man on the environment a key concern. With the advent of two world wars and the nuclear arms race, Americans entered a new realm of technological Armageddon musings. The MAD world of the nuclear age and projects such as Doom Town added technological anxiety to the mix. In 1938, a radio broadcast of H. G. Wells's *The War of the Worlds* threw Americans into disarray, with an alien attack seemingly on the horizon. The event demonstrated the power of modern media to manufacture catastrophe culture. In 1998, *Without Warning* (with meteors) reprised the broadcast, to notably less effect.

Environmental catastrophe culture also reflected the mood

of the moment. We think with the dominant media of time. In the late twentieth century and early twenty-first century, film and fiction provided a wealth of disaster imagery for consumption, as in movies such as *2012*. The Internet generation pointed and clicked on stories of the War on Terror, millennialism and global warming, spawning all manner of imaginative doomsdays. Cyberspace encouraged the blurring of real and virtual disaster scenarios.

The consumption of doomsday fears also proved businessworthy in the United States setting, interfacing with the direction of capitalism. Sales shifted from nuclear shelters in the 1950s and 1960s to cope with Doom Town to how to survive Armageddon in the 2000s. In 2009, Costco introduced THRIVE on to the marketplace, a package (or stockpile) of 5,011 food servings. The survivalist freeze-dried dehydrated food, enough for one person for one year, sold with a label advertising American 'Shelf-Reliance'.

Broad feelings of disassociation and powerlessness, of an anxious nation, resided in the catastrophe culture of the late twentieth and early twenty-first centuries. In standard psychology, 'catastrophizing' is a common extension of anxiety, the extrapolation and imagination of a worst-case scenario perpetuating fear. Situations become black and white, good or bad, artificially bipolar and extreme. Environmental disasters provided evidence for this culture, lending themselves to the imagination of 'bad' scenarios. In an anxious (or, as Elizabeth Wurtzel calls it, *Prozac Nation*), with record levels of antidepressant and anxiety medication, no wonder that environmental catastrophe gripped a people. Doomsday was located in rising rates of anxiety in the United States, an outgrowth of antidepressant culture. Doomsday anxiety provided a perfection of worst fears.

Environmental doomsday represented the ultimate anxiety, the worst Katrina-like storm. Many of the environmental disasters in *Inviting Doomsday* touched on fundamental fears of death, decay, and despoliation: they featured cognitive power. The French film *La Jetée* (1962), capturing a post-nuclear war experiment to return (through mind/time travel) to 'the moment' of the end revealed this exceptionalism of disaster well. The human guinea pig in the film revisited moments in time, finding love in such bizarre places as a natural history museum (referred to as the Museum of Timeless Animals). The film connected a death fixation of the individual with the masses. The quest to find and capture a specific moment in time – the 'endpoint' – is the sort of obsession to rival the Holy Grail. End-of-world fears remain powerful and poignant. They also maintain a story of nature within them. Nature features a key part in the dramaturgy of catastrophe culture. A nuclear war, a virus, an ecological disaster, global warming, population collapse, all have a 'nature' side to them. Few disasters have a non-nature dynamic.

Practically, catastrophe culture in the United States rested very much on the actual production of a series of disasters and failed responses to them. It rested in the disasters discussed here. Disasters fed catastrophe culture, strengthened images, and increased national anxiety. First-hand and mediated experiences of events such as Hurricane Katrina raised the fear factor. With modern media and 24-hour news, everybody absorbed on some level the disaster experience, by being there in actuality or vicariously and imaginatively. Past fictional and real disasters served as the references (or anchors) for the imagination of doomsday ahead. Objective case studies and intimate, traumatic experiences provided the underpinnings of disaster memory formation. Likewise, catastrophe culture made and reinforced the disasters, gave them form and substance. The social construction of disasters fused seamlessly with broader catastrophe culture, and perpetuated this disaster 'gaze'.

Finally, catastrophe culture blended ecological calamity with other doomsday scenarios: for example, Hurricane Katrina referenced themes of 9/11. Multi-catastrophe melded together a terrorist strike and a hurricane. The fact that emergency evacuation plans, insurance policies, and survival products fail to differentiate disasters furthered the synergy. The de-differentiation in the consumption of disaster imagery, the blending of events into one mega-event, make catastrophe culture all-enveloping. The film *2012* provided an example of this by throwing all disasters and fears together. Environmental disaster, on some level, seemed like all disasters.

Doomsday fixations: the downward spiral

Tracking this doomsday machine and monitoring catastrophe culture was the *Bulletin of the Atomic Scientists'* 'doomsday clock'. In 1953, as both the Soviet Union and the United States tested thermonuclear devices, and preparations for Doom Town began, the clock moved to two minutes to midnight. In 1991, with the end of the Cold War, the clock moved back 17 minutes. In 2007, the clock for the first time registered issues of peak oil, energy depletion and global warming, and moved forward in time. According to the *Bulletin*, environmental dangers in the twenty-first century signified the new doomsday on the horizon. The threat from nuclear MADness had mutated into concerns over global warming. Al Gore equally offered his own ticking doomsday clock in *An Inconvenient Truth*, noting 'We have just 10 years to avert a major catastrophe that could send our entire planet into a tail-spin'. Both Gore and the *Bulletin* sold environmental alarmism, informing their readers that time was running out, that humanity was spiralling downwards, and that the clock was ticking. They updated the Mayan calendar by offering new endpoints. In January 2012, the *Bulletin* clock moved one minute closer to doomsday, at five minutes. Spokesperson Lawrence Krauss explained: 'Faced with clear and present dangers of nuclear proliferation and climate change, and the need to find sustainable and safe sources of energy, world leaders are failing to change business as usual.'[6]

While the *Bulletin* clock keeps ticking, many of the disasters in *Inviting Doomsday* show an element of repetition, of a cyclical process marking the movement towards midnight. The environmental alarmism of *Silent Spring* became *An Inconvenient Truth*; the Santa Barbara oil spill morphed into the *Deepwater Horizon* spill; and *The Day After* was revised into *The Day After Tomorrow*. The cycle of doomsday imagination continued, the process remaining basically the same, just with a different view-finder card for each event. In fictive realms, it is easy to see the cycle at work. In the original *The Day the Earth Stood Still* (1951), alien Klaatu arrived on Earth to save humanity from destroying itself with atomic weapons. In the 2008 remake, Klaatu (Keanu Reeves) instead tackled global warming. Hollywood highlighted the 'replacement' system at work – one anxiety scenario with another, a different object of attention but the same cycle and fears on screen. The villains of dystopia at oil spills and chemical sites, the forgers of Armageddon, remain corporate bodies and distant government orchestrators. Chemical giant Monsanto appeared both in *Silent Spring* and in Disneyland.

We need to fuel our imaginations and our anxiety constantly. But this repeat cycle also threatens environmental numbness.

Films on repeat spell boredom. Environmental disasters on repeat provoke a failure to act. Alan Bryman described the process of de-differentiation of consumption in Disneyland, whereby the patron moved seamlessly from roller-coaster ride, to shopping mall, to fast food, to exit, with little cognition, or impact. A continual, even seamless, cycle of disasters can also inculcate little interest. American attention and anxiety demand new environmental dystopias, fresh hard-hitting incidents. As in *2012*, there is the need to increase the danger quotient, for the eco-disaster to seriously entertain. Science-fiction writer Robert Silverburg touched on this in 'When we Went to See the End of the World' (1972). People, offered a trip to see the end of the world, experience it as voyeurism, tourism, entertainment and speculation. Each of the travellers view something different, what they want to see. At the same time, in the background, real collapse happens about them without noticing. People find it impossible to differentiate between real and virtual, of the entertaining *2012* and actual disaster.[7]

Where does the series of disasters and doomsday scenarios mapped out here leave nature and the environment? What has the doomsday machine done to American nature? In the colonial era, settlers had limited means to transform land. But from a period of non-mastery, Americans emerged to enact massacres on the nineteenth-century frontier. The frontier doomsday machine devoured wilderness, killed it off. Nature moved from wild to tamed. People became disconnected from the natural world. In the twentieth century, DDT, genetic engineering and other technologies transformed nature further, damaged it beyond recognition. The bomb threatened wholesale collapse by radioactive fallout. Atomic doomsday seemed the perfect expression of detachment from the natural world by us bombing it to death. But, as the final chapter shows, new technologies offered the possibility of virtual nature, a cyber-world, and an end to the organic: a Disnature with no link to the material Earth but perfected in our image; a re-envisioning of nature based on entertainment and whimsy; a psycho-nature based on our psychology alone. Hence, through a series of events, nature has gone from wild, to tamed, before being transformed (technologically) then rendered bankrupt (virtually and with Disney). Inviting doomsday is the story of separation: making nature distant, remote and gone.

The series of disasters also reveals a downward spiral. They show extreme failures in government, corporations, and ultimately, humanity. Together, they forge a narrative of conquest, technological mishap, massacre, and ecological collapse. Despite protestations otherwise, we have yet to move past such a narrative. Katrina, global warming, oil spills and nuclear winter have similar dynamics. And our response has yet to catch up with the cycle of disaster. Norman Sanders related his concern over environmentalism, 'some people respond in the wrong ways. Many accept extinction as inevitable. Others discount the whole concept of environmental danger as sensational and exaggerated'. Despite messages that say otherwise, America is still inviting doomsday, and there are still only a few minutes to midnight.[8]

Notes

1. And concern over technology doing irretrievable damage, running amok with disastrous consequences, dates back at least to Mary Shelley's *Frankenstein* (1818).
2. P. D. Smith, *Doomsday Men: The Real Dr Strangelove and the Dream of the Superweapon* (Harmondsworth: Allen Lane, 2007).
3. Nicholas Thompson, 'Inside the Apocalyptic Doomsday Machine', *Wired*, 21 September 2009.
4. Lynn White, 'The Historical Roots of our Ecological Crisis', *Ecology and religion in history* (New York: Harper and Row, 1974).
5. See Karrie Craig et al., 'Environmental Factors in the Etiology of Anxiety', American College of Neuropsychopharmacology (2005).
6. Krauss: Jason Ukman, 'Doomsday Clock Ticks Closer to Midnight', *Washington Post*, 10 January 2012.
7. Robert Silverburg, 'When we Went to See the End of the World', (1972).
8. Norman Sanders, *Santa Barbara News Press*, 7 May 1970, MSS 11, 54, UCSB.

Epilogue: The Doomsday Seed

On the frozen Norwegian island of Spitsbergen, on the edge of the Arctic Circle, 620 miles (998 km) from the North Pole, 430 feet (131 m) above sea level, is an underground storage facility cut into the side of a mountain. Built at a cost of $9 million, it opened in 2008. Within it resides the doomsday seed.

Applauded as an international scheme with true environmental scope, the doomsday seed (officially titled Svalbard Global Seed Vault) set about tackling the age of global warming by preserving all of the seeds of the world. *Time* magazine called it a 'best invention' of 2008. Funders for the project included Bill Gates, the Rockefeller Institute (under the banner of the 'Global Crop Diversity Trust'), along with the Norwegian Government and Nordic Genetic Research Centre. The vault offered storage for 4.5 million seed samples. Within its first year, 400,000 were collected, entered cold storage and offered secure protection for potentially hundreds of years. In July 2008, President Carter, Ted Turner, Madeline Albright, Tom Dashle and Larry Page, founder of Google, all visited. United States congressmen donated chilli pepper seeds.[1]

The seed project had its share of historic precursors. In the 1940s, the United States established seed banks as one response to the Dust Bowl. In 1958, the US government set up the National Seed Storage Laboratory at Fort Collins, Colorado. Seed banks outside the United States included Wahlberg in Moravia and the

Soviet Pavlovsk seed bank in Vavilov, Leningrad, established in 1926. In 2004, the United Nations ratified the International Treaty on Plant Genetic Resources. The treaty encouraged a new 'global network' for seed collecting, banking and sharing that fed directly into the doomsday seed project.

The doomsday seed was also shaped by scientific and technological impulses, world capitalism and catastrophe culture. It epitomised the modern naturalist at work, collecting, annotating and preserving biota. A seeded version of the Linnaean system of classification, the bank collected as many varieties of life as possible. Critics of the scheme pointed to its commercial, rather than its botanical, uses, and raised the spectre of an agri-business conspiracy plot. Energy researcher F. William Engdahl fretted over the involvement of 'GMO companies, the so-called Four Horsemen of the Seeds Apocalypse – Monsanto, Syngenta, Dow Chemical and DuPont'. Motives of the scheme went questioned, with 'Using Climate Change' dubbed a 'pretext to Appropriate World Seeds' Treasure'. Rather than preserve wild seeds, cynics interpreted the storehouse as a ground zero for a new genetically modified future.[2]

Others saw the doomsday seed as a response to impending catastrophe. The *New York Times* described it as 'a sort of backup hard drive, in case natural disasters or human errors erase the seeds from the outside world'. *Wired* named it 'Nature's back up'. The facility was even designed to survive doomsday. As if by dress rehearsal, an earthquake of 6.2 on the Richter scale failed to register inside the bank.[3]

The project seemed to be the perfect product of environmental catastrophe culture. It signified a late reaction to disappearances and extinctions on an alarming scale. Scientists noted the huge number of plant species lost in earlier decades. Statistics included an 80 per cent loss in varieties of maize and 94 per cent in peas in the United States since the 1930s. Climate change threatened more pests, floods, and droughts. The vault offered refuge for seeds in the face of impending doomsday. One commentator declared it a 'repository of last resort for humanity's agricultural heritage'. For the *New York Times*, the doomsday seed was part of a 'broader effort to get plant data, to save nature while we can'.[4]

The doomsday seed also seemed an act of desperation, a last-ditch attempt to save the natural world. It equated to a twenty-first century update of Martha the passenger pigeon held in her cage. Fear drove the experiment. As Cary Fowler, President of the

Global Crop Diversity Trust, a non-profit group running the vault, explained:

> You need a system to conserve the variety so it doesn't go extinct . . . A farmer may make a bowl of porridge with the last seeds of a strain that is of no use to him, and then it's gone. And potentially those are exactly the genes we will need a decade later.

People needed nature to survive. It was also caught up in a culture of disaster. 'We started thinking about this post-9/11 and on the heels of Hurricane Katrina', said Fowler, 'Everyone was saying, why didn't anyone prepare for a hurricane before? We knew it was going to happen.'[5]

Visually, the seed bank most resembled a Cold War structure. Hidden in the side of a mountain, bomb proof and secure, the concrete carbuncle resembled a nuclear bunker. Like a Cold War structure, designed to survive attack, the doomsday seed featured high automation, high security (dubbed 'the Fort Knox of food') and could prosper well past humanity. No one person held the entry codes, and the structure featured 'security akin to a missile silo'. Requiring no permanent staff and working by automated temperatures and computers, the doomsday seed project appeared to be a machine ready to survive all kinds of calamity. The entrance jolted out of the mountain like a bit of skyscraper poking out of the snow, akin to the Statue of Liberty thrusting out of the sand in *The Planet of the Apes* (1968). The doomsday seed could survive Warm War (global warming) like a nuclear shelter could survive Cold War.[6]

The doomsday seed offered survival through doomsday. It promised new plants to cope with fresh climates ahead, proffer seeds suited to a post-Apocalyptic landscape. On some level, such 'seed survivalism' seemed straight out of Hollywood. But at least surviving doomsday had moved past a purely military angle. In the Cold War, survival technology was best seen in 'the doomsday plane' Boeing E4-B, an electro-magnetic pulse (EMP)-proof nuclear flyer, designed to overlook Armageddon and the atomic battlefield. Now survival meant ecology. Some saw the doomsday seed project as a twenty-first century Noah's Ark. As Norway's prime minister, Jens Stoltenberg, enthused, 'It is the "Noah's Ark" for securing biological diversity for future generations'. *National Geographic* declared, 'This is a frozen Garden of Eden', while the European Commission's president, José Barroso, told of 'a high-tech facility that could save the world'. Within the underground facility rested the seeds of a new world.[7]

Natural revival?

> On this first day of a new century we humbly beg forgiveness and dedicate these last forests of our once beautiful nation to the hope that they will one day return and grace our foul earth. Until that day may God bless these gardens and the brave men who care for them.
>
> Anderson, *Silent Running* (1972)

The doomsday seed smacks of desperate measures but at least continues an American interest in biodome experiments and nature preservation. The science-fiction film *Silent Running* (1972), a Hollywood response to eco-issues of overdevelopment and resource collapse, explored this enduring quest to preserve ecology

Figure E.1 *Silent Running* (Universal, 1972).

inside time capsules and laboratories. With biotic life extinguished on Planet Earth (akin to 'A Fable for Tomorrow' of *Silent Spring*), astronaut Freeman Lowell (Bruce Dern) cares for a giant greenhouse on board the spaceship *Valley Forge*. Along with three other vessels, the *Valley Forge* carries all that remains of nature, the last wilderness preserved in the last frontier of space. But Lowell rebels when presented with the order to destroy the biodome on board so that the *Forge* can be returned to commercial use. Robots join him in resistance, and Lowell ultimately detonates *Valley Force* with nuclear devices. He sacrifices himself and the ship, so that the biodome can float away freely. The last of nature survives, tended by a robot. The wilderness lives on, minus humanity.

With the doomsday seed, the argument follows that we need nature to prosper in order to survive doomsday. But what has actually happened at prototype doomsday landscapes? Across many of the disasters discussed in this book, the story is one of human and environmental loss in the short term but also recovery in the long term: at New Orleans, on the Great Plains, with DDT-affected birds, and across oil spills nature fights back. At Nevada Test Site, home of Doom Town, and popularly conceived as one of the most contaminated places on the planet, golden eagles rest at bombed-out craters. Nature has prospered in the nuclear wilderness, adapted, re-wilded and recolonised. Equally, Hanford Reach, once part of the Manhattan Project, became a National Monument in 2002. Of course, nature at such places may not need us.[8]

In *2012*, military personnel and scientists rushed 'couples' of animals to the last ships of humanity to join the rich and powerful capitalists on board. The ships served as four giant Noah's Arks. According to the film, the Earth survives its core overheating. According to many prophecies, better days follow the End of Days, and nature revives past doomsday. Notably in *2012*, however, the Point of Good Hope lives on but America itself disappears for good.

Notes

1. 'Best Inventions of 2008', *Time*, 29 October 2008.
2. F. W. Engdahl, 'Nato's Doomsday Seed Vault', *International News*, 22 September 2008.
3. Elizabeth Rosenthal, 'Near Arctic, Seed Vault is a Fort Knox of Food', *New York Times*, 29 February 2008; 'Nature's back up: The Global Seed Vault', *Wired*, 12 October 2010.
4. Engdahl (2008); Rosenthal (2008).

5. Rosenthal (2008).
6. Rosenthal (2008).
7. Doug Mellgren, 'Norway marks seed vault opening', *National Geographic*, 26 February 2008; John Roach, 'Doomsday Vault will End Crop Extinction', *National Geographic*, 27 December 2007.
8. John Wills, '"Welcome to the Atomic Park": American Nuclear Landscapes and the "Unnaturally Natural"', *Environment and History* 7/4 (November 2001).

Index